Praise for *Reinventing the Sacred*

"[*Reinventing the Sacred*] sparkles from every angle as its author gallops through the relevant science, philosophy, economics, history, ethics, poetry, and—well, we had better use the word because Kauffman does: religion. . . . Bringing science and religion together globally in the way that Kauffman wishes is not going to be easy—as other ecumenical movements have repeatedly found—but it is necessary."   —*Science*

"Drawing on his extensive grasp of leading-edge thinking across a wide array of sciences, Kauffman makes a convincing case for the inherent irreducibility of fundamental aspects of biology, psychology, and even physics itself. . . . Well-written and rigorously argued. . . . For this meaningful contribution to the quest for an era of sustainability, atheists and believers alike should be most grateful."

—*Shift: At the Frontiers of Consciousness*

"Kauffman, an outstanding thinker who has devoted much reflection to complexity theory, offers some insightful perspectives on the physical world. . . . This is an interesting book that will generate much discussion."

—*Choice*

"[Kauffman] offers a fresh angle in the ongoing debates concerning creationism, intelligent design, and evolution."   —*Library Journal*

"Provocative. . . . Kauffman raises important questions about the self-organizing potential of natural systems that deserve serious consideration."

—*Publishers Weekly*

"This brilliantly argued book takes science into novel territory with clarity and conviction, and in Kauffman's inimitable style it challenges some scientific taboos. With this book a new biology is emerging, and with it a new culture."

—Brian Goodwin, co-author of *Signs of Life:
How Complexity Pervades Biology*

"*Reinventing the Sacred* is a tour de force and a brilliant manifesto for a new emergence-based scientific worldview. But science alone will never be enough; humanity must also invent new categories of the sacred that speak to this naturalistic age. Stuart Kauffman courageously challenges fundamentalist pretensions on both sides, seeking to mold a new partnership of science and religious values . . . an epoch-making book."

—Philip Clayton, author of *Mind and Emergence*

"Stuart Kauffman has long studied the nature of complexity in biological systems. His new book shows in a startling way the power of these ideas in our understanding of ourselves and how we relate to the world around us. The sense of agency and of values, seemingly banished by the scientific viewpoint, are restored and enriched by a fuller perception of science deriving from biology as well as physics. Any reader's views will be dramatically altered."

—Kenneth Arrow, Nobel Laureate in Economics

"This is a brilliant, new, comprehensive, scientific world picture, and it deserves a wide reading in the educated public."

—Gordon D. Kaufman, Mallinckrodt Professor of Divinity, Emeritus, Harvard University

REINVENTING
THE SACRED

ALSO BY STUART A. KAUFFMAN

Investigations

At Home in the Universe

The Origins of Order

REINVENTING
THE SACRED

A New View of Science, Reason, and Religion

STUART A. KAUFFMAN

BASIC
BOOKS

A Member of the Perseus Books Group
New York

Books published by Basic Books are available at special discounts for bulk purchases in the
United States by corporations, institutions, and other organizations. For more information,
please contact the Special Markets Department at the Perseus Books Group, 2300 Chestnut
Street, Suite 200, Philadelphia, PA 19103, or call (800) 810-4145, ext. 5000, or e-mail
special.markets@perseusbooks.com.

Designed by Linda Harper

The Library of Congress has catalogued the hardcover as follows:
Kauffman, Stuart A.
Reinventing the sacred: a new view of science, reason, and religion / Stuart A. Kauffman.
    p.cm.
  Includes bibliographical references and index.
  ISBN 978-0-465-00300-6 (alk. paper)
  1. Complexity (Philosophy) 2. Religion and science. 3. Reductionism. 4. Life sciences—
Philosophy. I. Title.

Q175.32.C65K38 2008
215--dc22

                                                                          2007052263

Paperback ISBN 978-0-465-01888-8

To the conversations we must have

CONTENTS

Preface

The title of this book, *Reinventing the Sacred*, states its aim. I will present a new view of a fully natural God and of the sacred, based on a new, emerging scientific worldview. This new worldview reaches further than science itself and invites a new view of God, the sacred, and ourselves—ultimately including our science, art, ethics, politics, and spirituality. My field of research, complexity theory, is leading toward the reintegration of science with the ancient Greek ideal of the good life, well lived. It is not some tortured interpretation of fundamentally lifeless facts that prompts me to say this; the science itself compels it.

This is not the outlook science has presented up to now. Our current scientific worldview, derived from Galileo, Newton, and their followers, is the foundation of modern secular society, itself the child of the Enlightenment. At base, our contemporary perspective is reductionist: all phenomena are ultimately to be explained in terms of the interactions of fundamental particles. Perhaps the simplest statement of reductionism is due to Simon Pierre Laplace early in the nineteenth century, who said that a sufficient intelligence, if given the positions and velocities of all the particles in the universe, could compute the universe's entire future and past. As Nobel laureate physicist Stephen Weinberg famously says, "All the explanatory arrows point downward, from societies to people, to organs, to cells, to biochemistry, to chemistry, and ultimately to physics." Weinberg also says, "The more we know of the universe, the more meaningless it appears."

Reductionism has led to very powerful science. One has only to think of Einstein's general relativity and the current standard model in quantum physics, the twin pillars of twentieth century physics. Molecular biology is a product of reductionism, as is the Human Genome Project.

But Laplace's particles in motion allow only *happenings*. There are no meanings, no values, no doings. The reductionist worldview led the existentialists in the mid-twentieth century to try to find value in an absurd, meaningless universe, in our human choices. But to the reductionist, the existentialists' arguments are as void as the spacetime in which their particles move. Our human choices, made by ourselves as human agents, are still, when the full science shall have been done, mere happenings, ultimately to be explained by physics.

In this book I will demonstrate the inadequacy of reductionism. Even major physicists now doubt its full legitimacy. I shall show that biology and its evolution cannot be reduced to physics alone but stand in their own right. Life, and with it agency, came naturally to exist in the universe. With agency came values, meaning, and doing, all of which are as real in the universe as particles in motion. "Real" here has a particular meaning: while life, agency, value, and doing presumably have physical explanations in any specific organism, *the evolutionary emergence of these cannot be derived from or reduced to physics alone.* Thus, life, agency, value, and doing are real in the universe. This stance is called emergence. Weinberg notwithstanding, there are explanatory arrows in the universe that do not point downward. A couple in love walking along the banks of the Seine are, in real fact, a couple in love walking along the banks of the Seine, not mere particles in motion. More, all this came to exist without our need to call upon a Creator God.

Emergence is therefore a major part of the new scientific worldview. Emergence says that, while no laws of physics are violated, life in the biosphere, the evolution of the biosphere, the fullness of our human historicity, and our practical everyday worlds are also real, are not reducible to physics nor explicable from it, and are central to our lives. Emergence, already both contentious and transformative, is but one part of the new scientific worldview I shall discuss.

Even deeper than emergence and its challenge to reductionism in this new scientific worldview is what I shall call breaking the Galilean spell. Galileo rolled balls down incline planes and showed that the distance traveled varied as the square of the time elapsed. From this he obtained a universal law of motion. Newton followed with his *Principia*, setting the stage for all of modern science. With these triumphs, the Western world came to the view that all that happens in the universe is governed by natural law. Indeed, this is the heart of reductionism. Another Nobel laureate physicist, Murray Gell-Mann, has defined a natural law as a compressed description, available beforehand, of the regularities of a phenomenon. The Galilean spell that has driven so much science is the faith that all aspects of the natural world can be described by such laws. Perhaps the most radical scientific claim I shall make is that we can and must break the Galilean spell. I will show that the evolution of the biosphere, human economic life, and human history are partially indescribable by natural law. This claim flies in the face of our settled convictions since Galileo, Newton, and the Enlightenment.

If no natural law suffices to describe the evolution of the biosphere, of technological evolution, of human history, what replaces it? In its place is a wondrous radical creativity without a supernatural Creator. Look out your window at the life teeming about you. All that has been going on is that the sun has been shining on the earth for some 5 billion years. Life is about 3.8 billion years old. The vast tangled bank of life, as Darwin phrased it, arose all on its own. This web of life, the most complex system we know of in the universe, breaks no law of physics, yet is partially lawless, ceaselessly creative. So, too, are human history and human lives. This creativity is stunning, awesome, and worthy of reverence. One view of God is that God is our chosen name for the ceaseless creativity in the natural universe, biosphere, and human cultures.

Because of this ceaseless creativity, we typically do not and cannot know what will happen. We live our lives forward, as Kierkegaard said. We live as if we knew, as Nietzsche said. We live our lives forward into mystery, and do so with faith and courage, for that is the mandate of life itself. But the fact that we must live our lives forward into a ceaseless creativity that we cannot fully understand means that *reason alone* is an insufficient guide to living our lives. Reason, the center of the Enlightenment, is but one of the evolved, fully human means we use to live our lives. Reason itself has finally led us to see the inadequacy of reason. We must therefore reunite our full humanity. We must see ourselves whole, living in a creative world we can never fully know. The Enlightenment's reliance on reason is too narrow a view of how we flourish or flounder. It is important to the Western Hebraic-Hellenic tradition that the ancient Greeks relied preeminently on reason to seek, with Plato, the True, the Good, and the Beautiful. The ancient Jews, living with their God, relied more broadly on their full humanity.

The ancient Jews and Greeks split the ancient Western world. The Jews, as Paul Johnson wrote in his *History of the Jews*, were the best historians of the ancient world, stubbornly commemorating the situated history of a people and their universal, single God, our Abrahamic God. With this part of our Western Hebraic-Hellenic tradition comes our Western sense of history and progress, alive in the creativity of human history. In contrast, Greek thought was universalist and sought natural laws. The Greeks were the first scientists in the West.

If both natural law and ceaseless creativity partially beyond natural law are necessary for understanding our world, and if we as whole human beings live in this real world of law and unknowable creativity, these two ancient strands of Western civilization can reunite in ways we cannot foresee. Out of this union can arise a healing of the long split between science and the humanities, and the schism between pure reason and practical life, both subjects of interest to Immanuel Kant. Science is not, as Galileo claimed, the only pathway to truth. History, the situated richness of the humanities, and the law are true as well, as we will see. This potential union invites a fuller understanding of ourselves creating our histories and our sacred, as we create our lives.

Across our globe, about half of us believe in a Creator God. Some billions of us believe in an Abrahamic supernatural God, and some in the ancient Hindu gods. Wisdom traditions such as Buddhism often have no gods. About a billion of us are secular but bereft of our spirituality and reduced to being materialist consumers in a secular society. If we the secular hold to anything it is to "humanism." But humanism, in a narrow sense, is too thin to nourish us as human agents in the vast universe we partially cocreate. I believe we need a domain for our lives as wide as reality. If half of us believe in a supernatural God, science will not disprove that belief.

We need a place for our spirituality, and a Creator God is one such place. I hold that it is we who have invented God, to serve as our most powerful symbol. It is our choice how wisely to use our own symbol to orient our lives and our civilizations. I believe we can reinvent the sacred. We can invent a global ethic, in a shared space, safe to all of us, with one view of God as the natural creativity in the universe.

1

BEYOND REDUCTIONISM

Batter my heart, three-person'd God; for you
As yet but knock, breathe, shine, and seek to mend;
That I may rise, and stand, o'erthrow me and bend
Your force, to break, blow, burn and make me new.
I, like an usurpt town, to another due,
Labour to admit you, but Oh, to no end,
Reason your viceroy in me, me should defend,
But is captiv'd, and proves weak or untrue.
Yet dearly I love you, and would be loved fain,
But am betroth'd unto your enemy:
Divorce me, untie, or break that knot again,
Take me to you, imprison me, for I
Except you enthrall me, never shall be free,
Nor ever chaste, except you ravish me.

John Donne's exquisite "Holy Sonnet XIV: Batter My Heart," written in
about 1615, when he was a High Anglican churchman, speaks to one of

the most poignant schisms in Western society, and more broadly in the world: that between faith and reason. Donne wrote in the time of Kepler. Within a hundred years Newton had given us his three laws of motion and universal gravitation, uniting rest and motion, earth and the heavens: the foundations of modern science. With Descartes, Galileo, Newton, and Laplace, reductionism began and continued its 350-year reign. Over the ensuing centuries, science and the Enlightenment have given birth to secular society. Reductionistic physics has emerged for many as the gold standard for learning about the world. In turn, the growth of science has driven a wedge between faith and reason. It was not so much Galileo's geocentric theory (derived from Copernicus) that underlay his clash with the church but his claim that only science, not revelation, is the path to knowledge.

Today the schism between faith and reason finds voice in the sometimes vehement disagreements between Christian or Islamic fundamentalists, who believe in a transcendent Creator God, and agnostic and atheist "secular humanists" who do not believe in a transcendent God. These divergent beliefs are profoundly held. Our senses of the sacred have been with us for thousands of years, at least from the presumptive earth goddess of Europe ten thousand years ago, through the Egyptian, Greek, Abrahamic, Aztec, Mayan, Incan, and Hindu gods, Buddhism, Taoism, and other traditions. Neanderthals buried their dead. Perhaps they also worshiped gods. Recently an aboriginal tribe was unwilling to allow its DNA to be sampled as part of a worldwide study on the origins and evolution of humanity for fear that science would challenge its view of its own sacred origins. Ways of life hang in the balance. This book hopes to address this schism in a new way.

Part of my goal is to discuss newly discovered limitations to the reductionism that has dominated Western science at least since Galileo and Newton but leaves us in a meaningless world of facts devoid of values. In its place I will propose a worldview beyond reductionism, in which we are members of a universe of ceaseless creativity in which life, agency, meaning, value, consciousness, and the full richness of human action have emerged. But even beyond this emergence, we will find grounds to radically alter our understanding of what science itself appears able to tell us.

Science cannot foretell the evolution of the biosphere, of human technologies, or of human culture or history. A central implication of this new worldview is that we are co-creators of a universe, biosphere, and culture of endlessly novel creativity.

The reductionism derived from Galileo and his successors ultimately views reality as particles (or strings) in motion in space. Contemporary physics has two broad theories. The first is Einstein's general relativity, which concerns spacetime and matter and how the two interact such that matter curves space, and curved space "tells" matter how to move. The second is the standard model of particle physics, based on fundamental subatomic particles such as quarks, which are bound to one another by gluons and which make up the complex subatomic particles that then comprise such familiar particles as protons and neutrons, atoms, molecules, and so on. Reductionism in its strongest form holds that all the rest of reality, from organisms to a couple in love on the banks of the Seine, is ultimately nothing but particles or strings in motion. It also holds that, in the end, when the science is done, the explanations for higher-order entities are to be found in lower-order entities. Societies are to be explained by laws about people, they in turn by laws about organs, then about cells, then about biochemistry, chemistry, and finally physics and particle physics.

This worldview has dominated our thinking since Newton's time. I will try to show that reductionism alone is not adequate, either as a way of doing science or as a way of understanding reality. It turns out that biological evolution by Darwin's heritable variation and natural selection cannot be "reduced" to physics alone. It is emergent in two senses. The first is epistemological, meaning that we cannot from physics deduce upwards to the evolution of the biosphere. The second is ontological, concerning what entities are real in the universe. For the reductionist, only particles in motion are ontologically real entities. Everything else is to be explained by different complexities of particles in motion, hence are not real in their own ontological right. But organisms, whose evolution of organization of structures and processes, such as the human heart, cannot be deduced from physics, have causal powers of their own, and therefore are emergent real entities in the universe. So, too, are the biosphere, the human economy, human culture, human action.

We often turn to a Creator God to explain the existence of life. I will spend several chapters discussing current work on the natural origin of life, where rapid progress is being made. Self-reproducing molecules have already been demonstrated in experiments. A Creator God is not needed for the origin of life. More, you and I are agents; we act on our own behalf; we do things. In physics, there are only happenings, no doings. Agency has emerged in evolution and cannot be deduced by physics. With agency come meaning and value. We are beyond reductionist nihilism with respect to values in a world of fact. Values exist for organisms, certainly for human organisms and higher animals, and perhaps far lower on the evolutionary scale. So the new scientific view of emergence brings with it a place for meaning, doing, and value.

Further, the biosphere is a co-constructing emergent whole that evolves persistently. Organisms and the abiotic world create niches for new organisms, in an ongoing open textured exploration of possible organisms. I will discuss the physical basis of this "open texture" in the chapter on the nonergodic universe.

At a still higher level, the human economy cannot be reduced to physics. The way the diversity of the economy has grown from perhaps a hundred to a thousand goods and services fifty thousand years ago to tens of billions of goods and services today, in what I call an expanding economic web, depends on the very structure of that web, how it creates new economic niches for ever new goods and services that drive economic growth. This growth in turn drives the further expansion of the web itself by the persistent invention of still newer goods and services. Like the biosphere, the global economy is a self-consistently co-constructing, ever evolving, emergent whole. All these phenomena are beyond physics and not reducible to it.

Then there is the brute fact that we humans (at least) are conscious. We have experiences. We do not understand consciousness yet. There is no doubt that it is real in humans and presumably among many animals. No one knows the basis of it. I will advance a scientifically improbable, but possible, and philosophically interesting hypothesis about consciousness that is, ultimately, testable. Whatever its source, consciousness is emergent and a real feature of the universe.

All of the above speaks to an emergence not reducible to physics. Thus our common intuition that the origin of life, agency, meaning, value, doing, economic activity, and consciousness are beyond reduction to physics can be given scientific meaning. We live in a different universe from that envisioned by reductionism. This book describes a scientific worldview that embraces the reality of emergence.

The evolution of the universe, biosphere, the human economy, human culture, and human action is profoundly *creative*. It will take some detailed exploration of what are called Darwinian preadaptations to explain this clearly. The upshot is that we do not know beforehand what adaptations may arise in the evolution of the biosphere. Nor do we know beforehand many of the economic evolutions that will arise. No one foresaw the Internet in 1920. This unpredictability may exist on many levels that we can investigate. For example, we do not know beforehand what will arise even in the evolution of cosmic grains of dust that grow by aggregation and chemical reactions to form planetesimals. The wondrous diversity of life out your window evolved in ways that largely could not be foretold. So, too, has the human economy in the past fifty thousand years, as well as human culture and law. They are not only emergent but radically unpredictable. We cannot even prestate the possibilities that may arise, let alone predict the probabilities of their occurrence.

This incapacity to foresee has profound implications. In the physicist Murray Gell-Mann's definition, a "natural law" is a compact description beforehand of the regularities of a process. But if we cannot even prestate the possibilities, then no compact descriptions of these processes beforehand can exist. *These phenomena, then, appear to be partially beyond natural law itself.* This means something astonishing and powerfully liberating. We live in a universe, biosphere, and human culture that are not only emergent but radically creative. We live in a world whose unfoldings we often cannot prevision, prestate, or predict—a world of explosive creativity on all sides. This is a central part of the new scientific worldview.

Let me pause to explain just how radical this view is. My claim is not simply that we lack sufficient knowledge or wisdom to predict the future evolution of the biosphere, economy, or human culture. It is that these things are *inherently* beyond prediction. Not even the most powerful computer imaginable can make a compact description in advance of the regularities of these processes. There is no such description beforehand. Thus the very concept of a natural law is inadequate for much of reality. When I first discuss this in detail, in chapter 10, concerning Darwinian preadaptations, I will lay out the grounds for believing that this radical new view is correct. If it is, it challenges what I will call the Galilean spell, the belief that all in the universe unfolds under natural law.

There is a further profound implication: If the biosphere and the global economy are examples of self-consistently co-constructing wholes, and at the same time, parts of these processes are not sufficiently described by natural law, we confront something amazing. Without sufficient law, without central direction, the biosphere literally constructs itself and evolves, using sunlight and other sources of free energy, and remains a co-herent whole even as it diversifies, and even as extinction events occur. The same is true of the global economy, as we shall discuss in chapter 10. Such a self-organized, but partially lawless, set of coupled processes stands unrecognized, and thus unseen, right before our eyes. We appear to need a new conceptual framework to see and say this, then to understand and orient ourselves in our ever creative world. We will find ourselves far beyond reductionism, indeed.

Is it, then, more amazing to think that an Abrahamic transcendent, omnipotent, omniscient God created everything around us, all that we participate in, in six days, or that it all arose with no transcendent Creator God, all on its own? I believe the latter is so stunning, so overwhelming, so worthy of awe, gratitude, and respect, that it is God enough for many of us. God, a fully natural God, is the very creativity in the universe. It is this view that I hope can be shared across all our religious traditions, embracing those like myself, who do not believe in a Creator God, as well as those who do. This view of God can be a shared religious and spiritual space for us all.

This view is not as great a departure from Abrahamic thought as we might suppose. Some Jesuit cosmologists look out into the vast universe and reason that God cannot know, from multiple possibilities, where life will arise. This Abrahamic God is neither omniscient nor omnipotent, although outside of space and time. Such a God is a Generator God who does not know or control what thereafter occurs in the universe. Such a view is not utterly different from one in which God is our honored name for the creativity in the natural universe itself.

THE FOUR INJURIES

It would be a sufficient task to unravel the implications of this new scientific worldview for our unity with nature and life. But the project before us appears to be even larger. T. S. Eliot once wrote that with Donne and the other metaphysical poets of the Elizabethan age, for the first time in the Western mind, a split arose between reason and our other human sensibilities. The anguish between faith and reason in Donne's "Holy Sonnet XIV" is but one of these emerging schisms. With the growth of science and the Enlightenment, the Western mind placed its faith in reason and subordinated the rest of our humanity, Eliot's "other sensibilities," the fullness of human life.

Almost without our noticing, our secular modern society suffers at least four injuries, which split our humanity down the center. These injuries are larger than the secular-versus-religious split in modern society. What the metaphysical poets began to split asunder—reason and the remaining human sensibilities—we must now attempt to reintegrate. This is also part of reinventing the sacred.

The first injury is the artificial division between science and the humanities. C. P. Snow wrote a famous essay in 1959, "The Two Cultures," in which he noted that the humanities were commonly revered as "high culture" while the sciences were considered second-class knowledge. Now their roles are reversed: on many university campuses, those who study the humanities are often made to feel like second-class citizens. Einstein or Shakespeare, we seem to believe, but not both in the same room. This split is a fracture down the middle of our integrated humanity.

I believe it is important that this view is wrong. Science itself is more limited by the un-prestatable, unpredictable creativity in the universe than we have realized, and, in any case, science is not the only path to knowledge and understanding. I shall show in this book that science cannot explain the intricate, context-dependent, creative, situated aspects of much of human action and invention, or the historicity that embraces and partially defines us. These, however, are just the domains of the humanities, from art and literature to history and law. Truth abides here, too.

A second injury derives from the reductionistic scientific worldview. Reductionism teaches us that, at its base, the real world we live in is a world of fact without values. Wolfgang Kohler, one of the founders of Gestalt psychology, wrote a mid-twentieth-century book hopefully entitled *The Place of Value in a World of Fact*, in which he struggled unsuccessfully with this issue. His efforts had no effect on reductionism and its claims. The French existentialist philosophers struggled with the same issue, the view that the real universe is devoid of values. Our lives are full of value and meaning, yet no single framework offers a secure place for these facets of our humanity to coexist with fundamental science. We need a worldview in which brute facts yield values, a way to derive ought from is, just the step that Scottish Enlightenment philosopher David Hume warned against. Agency, values, and "doing" did not come into being separately from the rest of existence; they are emergent in the evolution of the biosphere. We are the products of that evolution, and our values are real features of the universe.

A third injury is that agnostic and atheist "secular humanists" have been quietly taught that spirituality is foolish or, at best, questionable. Some secular humanists are spiritual but most are not. We are thus cut off from a deep aspect of our humanity. Humans have led intricate and meaningful spiritual lives for thousands of years, and many secular humanists are bereft of it. Reinventing the sacred as our response to the emergent creativity in the universe can open secular humanists to the legitimacy of their own spirituality.

The fourth injury is that all of us, whether we are secular or of faith, lack a global ethic. In part this is a result of the split, fostered by reductionism, between the world of fact and the world of values. We lack a shared worldwide framework of values that spans our traditions and our responsibilities to all of life, one another, and the planet. Secular humanists believe in fairness and the love of family and friends, and we place our faith in democracy. Our diverse religions have their diverse beliefs. But in the industrialized world all of us are largely reduced to consumers. It is telling that the Nobel laureate economist Kenneth Arrow, when asked to help evaluate the "value" of the U.S. national parks, was stymied because he could not compute the utility of these parks for U.S. consumers. Even in our lives in nature we are reduced to consumers, and our few remaining wild places, to commodities. But the value of these parks is life itself and our participation in it.

This materialism profoundly dismays many thoughtful believers in both the Islamic world and the West. The industrialized world is seen to be, and is, largely consumer oriented, materialistic, and commodified. How strange this world would seem to medieval Europe. How alien it seems to fundamentalist Muslims. We of the industrialized world forget that our current value system is only one of a range of choices. We desperately need a global ethic that is richer than our mere concern about ourselves as consumers. We need something like a new vision of Eden, not one that humanity has forever left but one we can move toward, knowing full well our propensities for both good and evil. We need a global ethic to undergird the global civilization that is emerging as our traditions evolve together.

Part of reinventing the sacred will be to heal these injuries—injuries that we hardly know we suffer. If we are members of a universe in which emergence and ceaseless creativity abound, if we take that creativity as a sense of God we can share, the resulting sense of the sacredness of all of life and the planet can help orient our lives beyond the consumerism and commodification the industrialized world now lives, heal the split between reason and faith, heal the split between science and the humanities, heal the want of spirituality, heal the wound derived from the false reductionist belief that we live in a world of fact without values, and help us jointly build a global ethic. These are what is at stake in finding a new scientific worldview that enables us to reinvent the sacred.

2

REDUCTIONISM

Our scientific worldview deeply affects our view of our place in the universe, and the worldview put forward by reductionist science has created a dilemma for many people of faith. Those who believe in a transcendent God, one who answers prayers and acts in the universe, find that their God must either become a God of the gaps, active only in the areas science has yet to explain, or must act in contravention of scientific expectations. Neither alternative is satisfactory. For secular humanists, the very reality we cleave to is largely based upon reductionism. This, for reasons I will explain, is also proving to be unsatisfactory.

What, then, is reductionism? This philosophy dominated our scientific worldview from the time of Descartes, Galileo, Kepler, and Newton to the time of Einstein, Schrödinger, and Francis Crick. Its spirit, still adhered to by the majority of scientists, is captured by the physicist Steven Weinberg's two famous dicta: "The explµanatory arrows always point downward" to physics, and "The more we comprehend the universe, the more pointless it seems." In brief, reductionism is the view that society is to be explained in terms of people, people in terms of organs, organs by cells, cells by biochemistry, biochemistry by chemistry, and chemistry by physics. To put it even more crudely, it is the view that in the end, all of reality is *nothing but* whatever is "down there" at the current base of physics: quarks or the famous strings of string theory, plus the interactions among these entities. Physics is held to be the basic science in terms of which all other sciences will ultimately be understood. As Weinberg puts it, all explanations of higher-level entities point down to physics. And in physics there are only happenings, only facts.

The reductionist world, where all that exist are the fundamental entities and their interactions, and there are only happenings, only facts, has no place for value. Yet we humans, who are presumably reducible to physics like everything else, are agents, able to act on our own behalf. But actions are "doings," not mere happenings. Moreover, agency creates values: we want certain events to happen and others not to happen. How can values and doings arise from particle interactions where only happenings occur? Nowhere in the reductionistic worldview does one find an account of the emergence and reality of agency in the universe, which I discuss in chapter 7. The eighteenth-century skeptic philosopher David Hume, without recognizing it explicitly, bases his view on reductionism without the emergence of agency when he reasoned that one cannot deduce "ought" from "is." This is the so-called naturalistic fallacy. That is, he would have said, from the fact that women give birth to infants it does not follow that mothers should love their children. In short, reasoned Hume, from what happens, we cannot deduce what ought to happen. There is no scientific basis for the reality of values in the reductionistic worldview. This feature of reductionism led the post–World War II French existentialist philosophers to say, in anticipation of Weinberg, that the universe is meaningless and thus absurd, and to seek values in the choices we make. Yet choices themselves presume an agency capable of making choices, an agency whose reality is denied by reductionism. Thus, again, here are only happenings, no doings, actions, values, or choices, in reductionism.

In later chapters I will attempt to lay out the scientific foundations for agency and therefore value in the biological world, and for the evolutionary origins of ethics and "ought." As we shall see, agency is both real and emergent and cannot be reduced to the mere happenings of physics. This will provide an answer to Hume, who rightly says we cannot deduce "ought" from "is," values from mere happenings. Values are part of the language appropriate to the nonreducible, real, emergent, activities of agents. Thus agency and value bring with them what philosophers call teleological language, that is, language involving a sense of purpose or "end," as in our common explanations for our actions based on our reasons and intentions. Teleological language has long been a contentious issue among scientists and philosophers, many of whom consider it unscientific. I strongly disagree. Agency is emergent and real, but not reducible to physics, I shall argue, because biology is not reducible to physics. The biosphere, I will argue, is laden with agency, value, and meaning. Human life, which is certainly laden with agency, value, and meaning, inherits these qualities from the biosphere of which it is a part.

To better understand reductionism we may start, interestingly, with Aristotle. Aristotle argued that scientific explanation consisted of deduction via syllogism, as in: All men are mortal. Socrates is a man. Therefore, Socrates is mortal. Aristotle thus gave Western culture a model of reasoning that began with universal statements of fact (all men are mortal), considered specific cases (Socrates is a man), and then deduced the conclusion by applying the universal rule to the specific case (Socrates is mortal).

Now consider Newton's laws of mechanics, which consist of three laws of motion and a law of universal gravitation. They conform beautifully to Aristotle's mandate that scientific explanation should begin with a universal statement and systematically apply this statement to specific cases.

A central feature of Newton's laws is that they are deterministic. Roughly, here is how they work: picture a billiard table with balls moving on it. The balls are confined in space by "boundary conditions," namely the cushions on the walls, the pockets, and the tabletop. At any instant, all the balls are moving in specific ways. Their centers of mass have precise positions, and they have precise directions and speeds of motion, or velocities. These initial and boundary conditions are analogous to the specific case, Socrates is a man. Newton's laws, applied to this trivial case, state that given the current positions and velocities of all the balls and the boundary conditions, it is possible to compute, i.e., deduce, the trajectory of each ball as it bounces off other balls or the walls, or falls into a pocket. If the balls were started again in exactly the same positions and with the same motions, precisely the same trajectory of each ball would recur. This precise recurrence is what we mean by determinism. The initial conditions and boundary conditions exactly determine the system's evolution.

The position and velocity of each ball can be represented, in general, by six numbers, three numbers for the position of the ball in three-dimensional space, plus three numbers for the velocity of the ball "projected" onto the three spatial axes. Thus six numbers specify the position and velocity of each ball. If there are N balls, then the combined positions and velocities (or momenta) of the N balls can be specified by $6N$ numbers in what is called a $6N$-dimensional "state space." In general, we cannot draw a picture of more than a three-dimensional space, but mathematically we can consider a dimensional space where each axis corresponds to one of the $6N$ numbers. In this $6N$-dimensional state space, each unique set of positions and velocities of all the balls on the table corresponds to a single point. The time evolution of the whole system can thus be represented as a single trajectory in this massively multidimensional-state space.

Determinism means that under Newton's laws of motion (and in the absence of outside forces), there is only one possible trajectory from any point in the state space. Newton's laws are wonderfully successful. We send rockets on complex voyages through the solar system guided by nothing else. But these laws raise a profound problem. One of their fundamental features is that they are time-reversible: if the motions of all the balls on our billiard table were exactly reversed, the same laws would apply, and the balls would precisely retrace backwards their former forward motions. Yet as Humpty Dumpty famously discovered, we are not time reversible. Neither is the world around us. This time reversibility of Newton's laws (and also the time-reversible laws of quantum mechanics, described in chapter 13) has given rise to the so-called problem of "time's arrow," the distinction between past and future. Many physicists consider the famous second law of thermodynamics, which states that disorder in the universe will always increase, to be the physical foundation of the arrow of time. As a crude start on the second law, imagine a drop of ink in a dish of water. In time it will diffuse out to a uniform distribution. But if you start with the uniform ink distribution, it will not spontaneously diffuse back to an ink drop. The forward direction of time looks different from the backward direction. But in either direction, the ink particles and water molecules all follow Newton's time-reversible laws. The second law hopes to explain why the time asymmetry, forward time versus backward time, arises despite the time reversibility of Newton's laws. (As we shall see, doubts are now arising about the explanation of the second law of classical thermodynamics from statistical mechanics.)

Within a century after Newton, Pierre Simon Laplace had generalized Newton's laws to consider an arbitrarily large set of masses, or particles. Realizing that the particles would all simultaneously follow the laws of motion, Laplace imagined a "demon" of unfathomable intelligence. If supplied with the instantaneous positions and velocities of all the particles in the universe, declared Laplace, this demon would know the entire future history of motion of all the particles, and also, thanks to the time-reversibility of Newton's laws, the entire past motions of the particles. In short, to a sufficient intelligence, the entire past and future of the universe, ourselves and our operas included, could be calculated from a precise statement of the present positions and velocities of all the universe's particles. Laplace wrote in a deeply religious era. When asked by Napoleon what place existed for God in his system, Laplace replied that he had no need for that hypothesis. We see here in stark terms the wedge science was driving between reason and faith.

Laplace's vision is perhaps the purest and simplest statement of reductionism. Here there are two features. First is determinism, which was later abandoned with astonishment when quantum mechanics began to emerge. Second is the assumption that the entire universe and all the events within it, from particles colliding to nations at war, could be understood as nothing but the motions of a very large number of particles. It is here that we find the foundation of the belief that all higher order processes in the universe, such as organisms evolving, are ultimately to be explained as nothing but particles in motion. Weinberg's explanatory arrows point ever downward.

Over the next century or so, the determinism of Newton's laws led to a transformation of religious belief. Much of educated Europe shifted its belief from a theistic God who intervened in the running of the universe, for example, to answer prayer, to a deistic God who created the universe, set the initial conditions and boundary conditions, and allowed Newton's laws to unfold deterministically. In this worldview, there is no place for God's intervention in the running of the universe, hence prayers are not answered. God made the universe but does not act thereafter in its running. So our scientific worldview does color our sense of our place in the universe, our theology, and, equally importantly, our understanding of ourselves.

With respect to human action, determinism led to a contentious debate about "free will" that continues unabated. In the deterministic view, we are machines, and free will is an illusion. One solution to this problem was the idea that the mind and the body are composed of two radically different stuffs. The philosopher René Descartes, who preceded Newton, proposed that humans are composed of a mechanical body, part of res extensa, and a separate mental portion, res cogitans—an idea known as dualism. Free will lay in res cogitans. In part, Descartes invented dualism to save free will from what with Descartes, Galileo, and later became Newtonian determinism in the material world. But dualism raised the question of how the mind acts on the body. Descartes opted for action via an organ in the brain, the pineal gland, which is a rather improbable hypothesis.

The deepest claim of reductionism is that all events in the universe, from asteroid collisions to a kiss to a court in France finding a man guilty of murder, are "nothing but" the motions of particles. As philosophers like to put it, the "furniture of the universe" is limited to Laplace's particles in motion. This "nothing but" view survived even after quantum mechanics did away with the determinism of Newton and Einstein.

Quantum mechanics fundamentally altered the deterministic world of classical physics. To state the matter very simply, on its standard Bohr, or "Copenhagen" interpretation, quantum mechanics replaced the billiard-table universe of Newton and Einstein with a strange new view of persistent possibilities governed by the Schrödinger equation. The Schrödinger equation, which is itself deterministic, resembles the equations for water waves propagating across a lake. Much as water waves have a height or amplitude at each point in the lake, the Schrödinger equation determines at each point in space and time a height, that is an amplitude, of the Schrödinger wave. Mathematically, the square of this amplitude is, under the Born rule, the probability that a specific quantum process, if measured by a classical apparatus, will occur—for example, that a photon will be polarized in a specific way. In the wonderful weird world of quantum mechanics, which applies to very small things such as atoms, *nothing actual* happens when the Schrödinger wave equation propagates its wave. *Everything remains a propagating wave of mere possibilities*, each of which has a probability of a corresponding event being observed if the event is measured. It is only when the event is measured by a big, or macroscopic, non-quantum measuring apparatus (in the famous Copenhagen interpretation of quantum mechanics) that an *actual, real, or classical event*, say a photon hitting a photographic detector array, *happens*. The connection between quantum uncertainty and classical actualities is a matter of scientific debate to this day, seventy years after Schrödinger invented his equation. But even though little remained of classical determinism in the world of quantum probabilities, the most important part of reductionism emerged stronger than ever: the world was still nothing but particles in motion.

Having outlived determinism, the "nothing but" view may soon outlive particles as well. Some physicists, namely string theorists, now doubt that the world is ultimately made of particles. Rather, they say, it is made of one-dimensional strings that vibrate, and the modes of vibration correspond to different particles. Other versions of string theory posit higher-dimensional "branes" that vibrate. If string theory should be correct, then reductionism would hold that what is real are strings and possibly two- or higher-dimensional "branes" in vibratory and other motions.

In short, while Newton's laws have been surpassed by general relativity and quantum mechanics, the firm reductionistic worldview is that, at bottom, there is nothing but whatever is "down there" at the base of physics, plus Einstein's spacetime. What we think of as "down there" has evolved as physics has grown from its original basis in mechanics to include electromagnetism and the electromagnetic field, to the discovery of the atom, to the discovery of the proton, neutron, and electron, with which all of us are familiar. Now we have the "standard model" of particle physics, which unifies three of the known forces: the electromagnetic force that underlies electric and magnetic phenomena, the weak force that governs radioactive decay, and the strong force that holds atomic nuclei together. Beyond the standard model, there is a possible unification of general relativity and quantum mechanics, the two pillars of twentieth-century physics, through string theory, which would bring the fourth force, gravity, into the same mathematical framework as the other three.

But reductionism remains in place. As Weinberg said, all explanatory arrows point downward, from societies to people, to organs, to cells, to biochemistry, to chemistry, to physics. What is real is physics and the physical interactions of whatever is down there in the molecules, stars, organisms, biospheres, and legal proceedings in law courts.

Many outstanding scientists, including Nobel laureate physicists Murray Gell-Mann, Stephen Weinberg, and David Gross, are reductionists. Theirs is the "nothing but" view. It is not foolish. Indeed, these physicists would strongly argue that reductionism is the only sensible approach to physics. They can buttress their argument with several centuries of frankly stunning successes.

But if all explanatory arrows point downward, it is something of a quiet scandal that physicists have largely given up trying to reason "upward" from the ultimate physical laws to larger-scale events in the universe. Those downward-pointing explanatory arrows should yield a crop of upward-pointing deductive arrows. Where are they? Trouble arises, my physicist friends tell me, even in trying to generate the well-known equations for the behavior of a liquid, the Navier-Stokes equations, from more fundamental quantum physics. No one knows how to get there from the quantum mechanics of fluid atoms or molecules. Most physicists will admit that as a practical matter, the Navier-Stokes equations cannot be deduced. But most believe that, in principle, they could be deduced. Ultimately this assertion is not so much scientific as aesthetic: it amounts to a faith that nothing is required beyond quantum mechanics to explain fluid behavior, even if there is no practical way to carry out the deductive inferences to get from quantum mechanics to a river. The fluid system is still nothing but particles in motion. Reductionism holds, claim the reductionists.

It is less clear, but probably true, that many physicists would hold to the "nothing but" view in even more challenging situations. The couple in love and the man found guilty of murder in France, they would say, are likewise nothing but a set of basic physical events and their (extremely) complex interactions.

Why does this matter to us? The reason is Weinberg's second famous dictum: "The more we comprehend the universe, the more pointless it seems." Once again, what about organisms evolving, values, doing, acting, meaning, history, and opera? Are these all not real, not part of the furniture of the universe? Is science to have nothing to say about it? To accept this is to resign oneself to an impoverished view of both science and the world. One can empathize with the reductionist philosophy. It seems so tough-minded and clearheaded. So much outstanding science has been accomplished under its influence and guidance. I empathize but do not agree at all that reductionism alone suffices to know the truth of the world, and more broadly to understand it. As we shall see in the next chapter, it is not even clear that the Navier-Stokes equations can be reduced to the standard particles in motion of physics. The physicists are beginning to doubt the adequacy of reductionism alone in a quiet rebellion little known outside of academic physics departments. This rebellion is an initial hint that we need to change our scientific worldview.

3

THE PHYSICISTS REBEL

World-class physicists have begun to question the adequacy of reduction even within physics itself. Perhaps the first to do so publicly was Nobel Laureate physicist Philip W. Anderson, who published a famous paper in *Science* in 1972 entitled "More Is Different."

"More Is Different" is based on the physicist's concept of symmetry breaking. Consider, for example, a vertical pole on a flat surface—say a pencil miraculously balanced on its eraser. Before it falls, physicists speak of the pencil as having full 360-degree symmetry. After it falls in some direction, say to the north, it has broken that full 360-degree symmetry of the plane surface. A specific direction has emerged. This is a simple example of symmetry breaking, in which the underlying physical symmetry of the plane is not violated, but the broken symmetry creates a new macroscopic condition: a pencil pointing north, from which other specific broken symmetries may flow.

A specific symmetry that Anderson considers concerns ammonia. Ammonia has one nitrogen atom attached to three hydrogen atoms. These form a tetrahedron, a four-sided pyramid in which the nitrogen is at the top and the three hydrogen atoms form the three vertices of the base.

Now there is an underlying symmetry of three-dimensional space. Given this symmetry, the nitrogen could also be below the three hydrogens, making an inverted tetrahedron. "By quantum mechanical tunneling," Anderson wrote, "the nitrogen does 'leak' through the triangle of hydrogens to the other side, turning the pyramid inside out, and can do so very rapidly. This is the so-called inversion, which occurs at a frequency of about 3 x 10 raised to the 10th power per second." Thus there is no stable state of the pyramid on any timescale much longer than the inversion frequency. But "more is different," says Anderson, who then moves on to more complex molecules such as glucose, a six-carbon sugar molecule synthesized in your cells, that can be left- or right-handed. This handedness is called chirality and derives from the fact that some atoms, such as carbon, can, for example, have four different atoms chemically bonded to them. Then this can occur in mirror-symmetric, left-handed, or right-handed ways. Many molecules in our body are chiral rather than an equal, or racemic, mixture of both left and right versions. The amino acids that make up proteins in us are left-handed. A glucose molecule could, in principle, also quantum tunnel from right- to left-handed, but will not do so during the life of the universe. So the fact that more atoms are involved, and that organisms enzymatically construct only right-handed sugars, means that left-right symmetry is broken in sugar molecules in cells. This symmetry breaking is an emergent phenomenon. It does not violate any laws of quantum mechanics, but the specific handedness of biological sugar molecules is literally a fact of life. It is akin to the pencil falling and pointing north, from which further symmetries may be broken. In neither case do the fundamental laws of physics tell us which way the symmetry will be broken. Thus, concludes Anderson, the way the symmetry is broken is emergent.

In criticizing reductionism as inadequate to account for chirality—the explanatory arrow here does not point to quantum mechanics—Anderson admits "reduction" but denies that the inverse "constructivist" program will work where the constructivist program is to deduce or explain higher-level phenomena on the basis of fundamental physics. "The constructivist hypothesis," he writes, breaks down when confronted with the twin difficulties of scale and complexity. The behavior of large and complex aggregates of elementary particles, it turns out, is not to be understood in terms of a simple extrapolation of the properties of a few particles. Instead, at each level of complexity entirely new properties appear, and the understanding of the new behaviors requires research which I think is as fundamental in its nature as any other. That is, it seems to me that one may array the sciences roughly linearly in a hierarchy, according to the idea: The elementary entities of science X obey the laws of science Y.

X	Y
Solid state or many-body physics	elementary particle physics
chemistry	many-body physics
molecular biology cell biology	chemistry
*	*
*	*
*	*
psychology	physiology
Social sciences	psychology

But this hierarchy does not imply that science X is "just applied Y." At each stage entirely new laws, concepts, and generalizations are necessary, requiring inspiration and creativity to just as great a degree as that in the previous one.

Anderson has, in private conversations, since introduced another argument for emergence. Consider, he said, a computer. We all know that digital computers typically manipulate 1 and 0 symbols, called bits. The mathematician Alan Turing showed that his Turing machine, invented to mechanize computation, was computationally universal. That is to say, every well-specified sequence of mathematical operations that might be carried out on an alphabet of symbols could be carried out on his machine. Such a calculation is called an algorithm. So an algorithm is an "effective procedure" to get from some input to some output. After John von Neumann added serial processing architecture to Turing's machine, the modern computer was born. Such a computer can be programmed to carry out an enormous variety of algorithmic calculations.

There are, however, wonderful puzzles such as the halting problem. Suppose you program a computer to compute a result, print it out, then halt, where the halting of the computer algorithm means that the answer has been computed. Alonzo Church proved that there can be no general algorithm, or program, that is capable of examining any other program and computing how long it will take to halt, or even saying whether it will halt in less than infinite time. This means that the only way to find out if a program halts is to run it and see. This result is related to Gödel's incompleteness theorem, which established an astonishing mathematical fact. Early in the twentieth century, the great mathematician David Hilbert had pursued a program of axiomatizing mathematics. This required that all mathematical statements that were true, given a set of axioms, must be deducible from those same axioms. This property of an axiom system is called completeness. Gödel forever changed mathematics in 1931 by proving that Hilbert's program cannot succeed. Even modestly rich axiom systems, such as that for arithmetic, have the property that there are mathematical statements that are true but that cannot be deduced from the system's axioms. Mathematical systems are typically not complete; they always contain some statements that are "formally undecidable." If mathematics is not complete, not reducible to theorems derived from any given set of axioms, it might be less surprising if, in other areas of science, not everything is reducible to physics. Anderson argued that computation is also not reducible to physics. Anderson's point is the following: any system of entities able to "carry" 1 and 0 symbols can be used to compute. Students at MIT regularly build Tinkertoy systems to compute specific problems. With more difficulty, I could specify 1 and 0 "symbols" as different levels of water in water buckets, get a bunch of graduate students to pour water back and forth according to rules I establish, and carry out a computation that way. The same computation can be carried out on the ENIAC, a monstrously large early computer with thousands of triodes connected into vast circuits. Or my laptop could do it.

Anderson pointed out that since mathematically identical computations could be carried out with water buckets or a silicon chip, the computation itself is fully independent of the physical instantiation of the machine that carries it out. Thus the computation cannot be reduced to any specific physics. This is what philosophers call the multiple-platform argument.

But what exactly does it mean to "reduce" a computation to the behavior of one or more physical systems? Philosophers have thought about this for a long time and their answer is fully cogent. Consider the description of some event or object, such as our hypothetical computation or the man found guilty of murder in a French court. A "reduction" from this description to a more "fundamental" physical description requires that the words in the higher-level description are, in effect, shorthand for more precise lower-level descriptions. This requires that a given statement in the higher-level "language" be replaceable by a prestated and finite set of statements in the more fundamental language that are "jointly necessary and sufficient" for the higher-level statement. "Jointly necessary and sufficient" just means that, together, these lower-level statements can be substituted for the higher-level description without altering the truth or falsity of the higher-level description.

With this definition of reduction in place, it is clear how Anderson's multiple-platform argument about the computer system works. Suppose we want to add up the first 100 digits of a large number, and the sum in this case is 550. Out of the infinite variety of physical computers that could be created, there is no prestatable, finite list of conditions that is jointly necessary and sufficient for the statement, "The computer computes that the sum of the first 100 digits is 550."

In colloquial language, Anderson is arguing that the computer's ability to compute this sum does not depend upon its specific physics. Thus the computation cannot be reduced to physics.

In thinking about reductionism, some philosophers have introduced a concept called supervenience. Here the *finite* prestated set of statements at the lower level is allowed to be *infinite*. In more detail, some philosophers of science have distinguished between "types"—general cases—and "tokens"—specific instances. In the present case, a specific algorithm would be a type, and each specific realization of it on a specific physical computing system would constitute a token. Then the multiple-platform argument becomes the statement that the type statement can have an infinite number of token statements. My own view is that, insofar as supervenience is supposed to be a weakened form of reductionism, it does not work, for there typically is, in principle, no orderly way to generate the infinite list of token statements in question. Another way of saying this is that there typically does not seem to be an algorithm, an effective procedure, to generate the infinite list of lower-level statements. Supervenience, therefore, typically cannot be used for anything in trying to reduce, in a weakened sense, a higher-level description to a lower-level language.

Let's be clear about what Anderson's argument does not say: it does not say that any laws of physics are violated. Physics is perfectly intact, it is merely that a determined physicist cannot come up with a list of physical events that are jointly necessary and sufficient for the water buckets, the silicon chip, or whatever else to compute the sum of the first 100 digits.

Nobel laureate Robert Laughlin, a solid-state physicist, has also argued that reductionism alone is insufficient in physics. Laughlin focuses on what might be called ontological emergence, the emergence of new phenomena in the universe. His classic example is the temperature of a volume of gas particles at equilibrium. From statistical mechanics, we know that the classical thermodynamic quantity, "temperature," corresponds to the average kinetic energy of the gas particles. The faster the particles are moving, on average, the hotter the gas. The first point, says Laughlin, is that it makes no sense to speak of the temperature of a single gas particle, only of a collection of particles. The property of temperature depends upon the collective properties of a set of particles. More critically, the larger the number of gas particles in equilibrium in the system, the more precise is the measurement of temperature. Thus, Laughlin argues, temperature arises as a "collective emergent" property of the gas system, a property that is not present in any of its constituent particles but only in the whole. The defining character of such emergent properties is that their measurement becomes more precise as particle number increases.

Laughlin gives multiple examples. For example, rigidity is not a property of a single iron atom but is a collective property of an iron bar. Again, the more iron atoms in the solid, the better we can measure rigidity.

A third argument for the insufficiency of reductionism comes from the esteemed physicist Leo Kadanoff. Kadanoff has shown that the Navier-Stokes equations for fluid flow, which cannot yet be derived from quantum mechanics, can be derived but from a physicist's toy world of a hexagonal lattice with particles moving on it and obeying some simple laws. This ultimately becomes another multiple-platform argument. The Navier-Stokes equations are presumably derivable from the laws of quantum mechanics, even if we can't fathom how to do it. But since the same equations can also be derived from a hexagonal toy world, what sense does it make to say that they are reduced to just one of them, namely quantum physics rather than the toy world? Laughlin speaks of laws of "organization." It would appear from Kadanoff's toy world that the Navier-Stokes equations reflect a law of organization on a higher level than fundamental physics. They are emergent at their own level. Thus, contra Weinberg, the explanatory arrows need not point downward to particle physics and string theory, but to Kadanoff's toy world.

These three examples all come from Nobel-caliber physicists who seriously doubt the adequacy of reductionism alone—without ever asserting that any physical law is violated. That major physicists like Weinberg and Anderson so strongly disagree tells us that even in physics, reductionism is in serious question.

The generally acclaimed reduction of classical thermodynamics to statistical mechanics is a fourth example. It appears that this famous and widely accepted reductionist achievement does not entirely work yet. Classical thermodynamics was invented by Sadi Carnot in the early nineteenth century in an effort to understand the maximum efficiency of heat engines like steam engines. Included among its laws is the famous second law of thermodynamics, which gave us the concept of entropy and which asserts that in any isolated thermodynamic system that does not exchange matter or energy with its environment, entropy is constant or increases. It cannot decrease. As I mentioned in the last chapter, this "one-way-ness" is taken by physicists to account for the arrow of time.

Boltzman, Gibbs, and other physicists in the nineteenth century invented statistical mechanics in an effort to explain Carnot's thermodynamics in terms of the motion of particles governed by Newton's laws of motion. The basics of statistical mechanics is fundamentally simple. The centerpiece of the theory with respect to the second law of classical thermodyamics, the one-way increase of entropy, and thus the explanation of the arrow of time, is that a system of particles will, by chance (formalized as something called the ergodic hypothesis, which we will encounter in a later chapter), evolve to more-probable "macrostates" from less-probable macrostates. Thus a droplet of ink on a petri plate will typically diffuse to a uniform distribution. It will not un-diffuse from a uniform distribution to reconstitute the ink drop.

To be a bit more precise, think of the petri dish divided mathematically into many tiny volumes. Each possible distribution of ink molecules among these tiny volumes is a microstate. Clearly there are many more microstates corresponding to an approximate equilibrium distribution of the ink through the petri plate than there are microstates corresponding to the ink droplet in the center of the plate. Thus the uniform ink distribution is more probable than the ordered ink-drop distribution. The greater the number of microstates corresponding to the diffuse "macrostate," the greater the *disorder* of that macrostate: the same macrostate, here the diffuse distribution, can result from very many different microstates. The entropy of a macrostate is the mathematical logarithm of the number of microstates in that macrostate. So the ink in the center of the petri plate has fewer microstates, hence lower entropy, and is less probable than the higher-entropy, diffuse distribution to which the ink drop evolves. In other words, by random collisions, the ink system flows from the less probable to more probable macrostate. This one-way flow is the statistical-mechanics version of the second law in classical thermodynamics, the one-way increase in the entropy of a closed system. In the generally accepted view, classical thermodynamics has been successfully reduced to the random collisions of many particles under Newton's laws of motion. Thus statistical mechanics has reduced the arrow of time to the statistical increase in entropy.

But the philosopher David Alpert has pointed out a problem with this wonderful story. Given the time reversibility of Newton's laws, if the system started as the improbable, highly ordered ink drop and Newton's equations were solved *backward in time, the drop would again diffuse across the petri plate from less-probable macrostates to more-probable macrostates.* Entropy would appear to increase with time running backwards! What happened to the forward arrow of time, as reduced from classical thermodynamics to statistical mechanics? Boltzman knew of this issue and apparently never solved this fundamental problem. In short, the one-way arrow of time of classical entropy remains problematic in statistical mechanics. If so, the reduction of classical thermodynamics to statistical mechanics remains incomplete, for we apparently still need an account of the one-way increase in entropy in classical thermodynamics.

One approach to solving this problem is cosmological. It suggests that the universe started, in the big bang, in a highly ordered initial state rather like our ink drop, and has become less ordered overall ever since. But this solution, even if true, does not yield the conclusion that statistical mechanics by itself gives an arrow of time. Thus, this centerpiece example of the success of reductionist reasoning does not appear to be a success of reductionism at all.

Yet a further example arises due to a fundamental problem concerning the so-called constants of nature. The well-established equations of the standard model of particle physics, plus those of general relativity, together require about twenty-three constants that have to be adjusted "by hand" to yield the universe we know. The difficulty here is twofold. First, nobody has a theory to explain why these constants have the values they do. Second and far worse, were the values very different, they would yield universes vastly different from ours. For example, our universe has a number of properties that we rightly consider essential, including the capacity to form stars and galaxies, form complex atoms in stellar events, and form planets that include a high diversity of atoms, in order to create the conditions for our kind of life. Were the constants of nature greatly different, the universe that results would form no atoms, stars, galaxies, complex chemicals, or life. (There could be other life forms, incomprehensible to us, some of which might think of themselves as alive. But we'll ignore this point for now.) Various calculations have been carried out about how improbable the set of values of the constants of nature must be so that life is probable. Physicist

Lee Smolin, in his book *The Life of the Cosmos*, estimates crudely that the constants must be tuned to within 1 divided by 10 raised to the 27th power to be within range of creating a life-friendly universe.

In short, the constants of nature must be tuned exquisitely finely for intelligent living things able to wonder about why the constants of nature have the values they do to exist at all. How this happened is a deep problem.

There are several candidate answers to this problem. One is that we shall find a "final theory" that precisely predicts our universe and its constants. No one knows how to do this, although some theorists in quantum gravity are working on it. (Quantum gravity is the effort to unite quantum mechanics and general relativity in a common theory.) Smolin himself advanced an idea about how to "tune" the constants of nature that does not require any grand unification but has testable consequences. He proposed that baby universes are born in black holes—entities of such enormous mass that light itself falls into them—and these baby universes have minor variations in their constants. Universes with lots of black holes will have lots of baby universes, so a kind of cosmic natural selection occurs in which universes with many black holes will outnumber those with fewer black holes. Smolin hopes that such "cosmically selected" universes will have the desired small range of values of the constants. This idea predicts that minor variations in the constants of nature would reduce the expected number of black holes, a prediction partially confirmed by calculations and comparison with data.

Smolin's view implies that there need to be multiple universes. Other physicists have reached similar conclusions. This is a radical postulate, for, as far as we know, we can have no direct evidence of other universes. This very lack of evidence bears on the issue of reducibility. In what sense of reduction are we explaining something if we must appeal to entities such as other universes that are, or certainly now seem to be, in principle, unobservable, unknowable?

Making use of the idea of multiple universes, Weinberg himself was the first to suggest that the very existence of life in our universe might explain the constants of nature via what is called the anthropic principle. The anthropic principle comes in strong and weak forms. Why are we living in a universe whose constants are mysteriously "tuned" to accommodate life? One possibility is that an outside agent, conveniently a Creator God, tuned the constants. Virtually all scientists concur that this is not science. This is called the strong anthropic principle.

In various forms, the weak anthropic principle states that there are very many pocket universes, with a random or nonrandom distribution of the constants of nature among them. Only some will give rise to life, and of these only a few will support intelligent life able to wonder why the constants of nature have the values they do. So the "tuning" problem disappears because only those universes with good values of the constants will have life to wonder why the values of the constants are good. A number of cosmologists and physicists take the weak anthropic principle reasonably seriously.

All this is, of course, fascinating physics in its own right. But what is the connection to the broad claims of reductionism? This brings us again to Weinberg's "dream of a final theory" in the sense of a single theory, elegant like Einstein's general relativity, which would be the reductive theory of everything including the constants of nature. If this reduction to a single law in a single universe is what is meant by an ultimate reduction, then many physicists are giving up on it. Instead, the weak anthropic principle and the hypothesis of multiple universes is gaining ground. The single final theory is giving way to the view that physical law and the physical constants may be in a sense local to our universe, and no law at all explains why the constants have the values they do in our universe.

Another revolt against reductionism is the work of physicist Leonard Susskind of Stanford University. String theory is the most promising current effort to unite quantum mechanics and general relativity. It is based on a wonderful idea, which is that particles are not points but open or closed strings, or more complex objects, and that their patterns of vibration correspond to all the particles and forces. The early hope, when the theory was first explored, was that there would be a single string theory describing our universe, just as Weinberg aspired to. At present, however, it appears that there are a mind-numbing 10^{500} alternative string theories. String theory is in further trouble. We think of a theory in physics as a set of equations, but no one knows how to write down the equations for string theory in complete detail. In short, as Smolin writes in *The Trouble with Physics*, we don't really yet have a string theory, we have instead mathematically brilliant whiffs of a theory. Susskind, on the other hand, argues that the potential for 10^{500} string theories is to be welcomed, for within a megaverse of 10^{500} pocket universes, each with a random choice of string theory laws, many will surely be life friendly.

The weak anthropic principle, with its possibility of multiple universes, raises troubling questions about how well our scientists are adhering to the fundamentals of science. If we are to postulate multiple universes yet can have no access to them and cannot confirm their existence, have we actually explained anything at all? Perhaps someday we will manage to find evidence of multiple universes. Until then, the weak anthropic principle seems to stand on shaky evidential grounds.

In summary, the physicists are in a quiet revolution. More and more physicists, like Anderson, Laughlin, and Kadanoff, now seriously doubt the adequacy of reductionism alone. I hope to show in this book that the intellectual handcuffs of a rigid reductionism and a meaningless universe need no longer confine us. We are coming to a new scientific worldview that reaches to emergence and to vast unpredictability and unending, ever new diversity and creativity that appear to be beyond natural law itself.

4

THE NONREDUCIBILITY
OF BIOLOGY TO PHYSICS

Can biology be reduced to physics? Are organisms a real part of the furniture of the universe? Or are they just particles in motion?

Some people, especially in the United States, reject Darwinism, natural selection, and evolution. For virtually all biologists and other scientists, the historical fact of evolution is as well established as anything in science. I ask those of you who doubt this to please suspend judgment while we explore the implications of Darwin's ideas. They are of central importance to the new scientific worldview that I wish to advocate. If we can find a "third way" between a meaningless reductionism and a transcendental Creator God, which preserves awe, reverence, and spirituality—and achieves much more—your temporary suspension of judgment may be worthwhile.

Charles Darwin was educated in the early-nineteenth-century British tradition of natural theology, which saw in the staggering "fit" of organisms to their environments signs of the divine hand that created all life. He originally intended to go into the ministry, which in those days offered a comfortable home for men interested in the natural sciences. But after his famous voyage on the HMS *Beagle* in the early 1830s he came to doubt natural theology. The famous story about Darwin is that after his return from his trip on the *Beagle*, he read Thomas Malthus's *An Essay on the Principle of Population*, in which Malthus noted that the amount of food that could be produced increased, at best, arithmetically, while populations increased geometrically. Inevitably, people would outstrip their agricultural resources.

Upon reading Malthus, Darwin wrote in his diary that at last he had gotten hold of a theory to work with. In the natural world, too, he realized, species would proliferate and outrun their resources. A natural competition would arise, in which some would survive while others would not. Now, if organisms had offspring with variations, and if those variations in turn could be inherited by their own offspring, the variants whose features were most conducive to survival in the current environment would, over time, become more common. Organisms whose features were less conducive to survival would grow rarer and eventually die out. This, in a nutshell, is Darwin's idea of evolution by natural selection.

I have spent decades muttering at Darwin that there may be powerful principles of self-organization at work in evolution as well, principles that Darwin knew nothing about and might well have delighted in. Still, he had one of the great ideas of human history. With it, the entire view of biological life changed. A fixed diversity of species with puzzling similarities became descent, from parent to offspring, with modification. Biologists, following Carl Linnaeus, had already organized the living world into roughly hierarchical groupings based on anatomical similarities: species, genera, families, and so on, up to the major kingdoms of life. Why should this scheme work? The biologists of the time had looked at the systematic similarities among different groups of organisms (for instance among dogs, wolves, and foxes) and argued that there were "natural kinds." The set of natural kinds was thought to be fixed and to reflect God's intention in creating species for humanity's use or instruction. With the theory of evolution by natural selection, all that changed. The hierarchy of similarities became understood as descent with modification. Species or genera that had branched from a last common ancestor relatively recently would be more similar than those that had diverged earlier. For instance, dogs are closely related to wolves but more distantly related to cats, with which they share fewer similarities, because the last common ancestor of dogs and wolves is more recent than the last common ancestor of dogs and cats.

This is not the place to adumbrate the many stunning successes of Darwin's theory of evolution. But with one sweeping idea he made sense of the geological record of fossils, the similarity of organisms on islands to those on nearby major land masses, and many other facts. This is the hallmark of outstanding science. I say this because many who believe in the Abrahamic God still deny evolution and attempt to justify their denial on scientific grounds. This is a fruitless exercise. If you deny evolution, you will be astonished to see how much of the rest of science you also have to give up, from physics to chemistry to geology to biology and beyond. Evolution as a historical fact does not stand isolated from the intellectual matrix of the rest of science but is an integral, interwoven part of the tapestry of our entire understanding of nature.

The real reason many religious people oppose evolution is not scientific at all, of course, but moral. They see evolution as an amoral doctrine that denigrates values they rightly hold dear. They fear that the ethical foundations of Western civilization will crumble if evolution is factually true. Later I will explain why this fear is misplaced and argue that, in fact, evolution is the first, but not the only, source of our morality.

There is, however, considerable doubt about the power and sufficiency of natural selection as the sole motor of evolution. While no serious biologist disagrees that it plays a major role, some argue for other sources of order as well. I will describe my own efforts to find additional laws of self-organization in chapter 8.

In any case, since Darwin's time, natural selection has become a poster child for reductionist thinking. The modern synthesis, brought about by Ernst Mayr, Theodosius Dobzhansky, and others in the 1930s, united evolutionary theory with genetics; with the rise of molecular biology in the 1950s, genes changed from theoretical entities to specific sequences of DNA. We can now draw causal arrows from complex traits and behaviors to interactions among genes, from genes to molecules, from molecules to chemistry, and on down.

Still, not every causal arrow points downward. I want to discuss two features of Darwinism that—though they violate no law of physics—are not reducible to physics but are epistemologically and ontologically emergent: natural selection itself, and the attribution of functions to parts of organisms as due to natural selection.

This is the first time I have mentioned epistemological and ontological emergence, so I'll pause to explain that epistemological emergence means an inability to deduce or infer the emergent higher-level phenomenon from underlying physics. Ontological emergence has to do with what constitutes a "real" entity in the universe: is a tiger a real entity, or nothing but particles in motion, as the reductionists would claim? If the tiger is a real entity in its own right, it is ontologically emergent with respect to the particles comprising it. I will try to show that a tiger is both epistemologically and ontologically emergent with respect to physics.

Consider first the human heart and the attribution of function. If Darwin were asked, "What is the function of the heart?" he would answer, "To pump blood." But hearts also make thumping heart sounds—ask your doctor—yet these sounds are not the heart's function. Roughly, Darwin's claim is that the causal consequence for which hearts were selected was their capacity to pump blood, which conferred on the organisms with hearts a selective advantage. Darwin would thus give a selective evolutionary explanation for why hearts exist.

Now, notice again that the function of the heart is to pump blood, not to make thumping sounds. So we reach an important insight: *the function of a part of an organism is typically a subset of its causal features.* This means that we cannot analyze or understand an organ's function except in the context of its entire life cycle in its selective environment, and in the context of the selective history underlying its evolution. We need to know some of the evolutionary history of the biosphere to say anything about evolved functions.

Could a reductionist physicist deduce or explain the function of the heart? No, because physicists cannot even deduce the Navier-Stokes equations for fluid behavior. But let us grant, for the sake of discussion, that the physicist could deduce all the features of the heart from string theory. Then what? The physicist would deduce virtually *all* the properties of my heart. She would then have no way whatsoever to pick out, from the entire set of the heart's properties, the pumping of blood as the causal feature that constitutes its function. I will expand on this carefully.

First, grant that the physicist can deduce all properties of the heart. What would she have to do to pick out the pumping of blood as the relevant causal property that constitutes the heart's function? (Note that functions in this biological sense do not exist in physics.) She would have to explain that hearts have come to exist in the universe because pumping blood has conferred a selective advantage on the lineages of organisms that have evolved them. To carry out this explanation the physicist would have to analyze the whole organism in its selective environment and give an evolutionary account of the selection of the heart, just as Darwin or any biologist would do. But to do so, the physicist must temporarily become a biologist. She has to analyze the fossil record, demonstrate the emergence of the heart over time, and tie it to an account of the fact that hearts were selected and actually came into existence because of this selective functionality. Again, the physicist could conceivably deduce all the properties of the heart, but would have no means whatsoever to pick out the pumping of blood as the relevant causal feature that accounts for its existence.

But there is more: pumping hearts have causal powers that have modified the biosphere, from specific proteins and other molecules, to the *selective evolution of the heart's organizational behaviors and anatomical features that confer on it its capacity to pump blood,* to more features available to animals that possess hearts. The physicist cannot explain the heart without accounting for those (selected) causal powers and giving an evolutionary account of the role the heart has played in the further evolution of the biosphere. (I stress that biologists do not claim that all features of organisms are selected—for example, heart sounds are probably not selected.)

Now, even if the physicist could not deduce the function of the heart, is the heart a real part of the furniture of the universe? If we ask whether there is anything in the heart that is not just, say, particles in motion, the answer is no. Of course the heart is made of particles and not some mystical stuff. In this sense, reductionism works. But the heart works by virtue of its *evolved structure and the organization of its processes.* I will have a great deal to say about such organized processes, for they are less well understood than you might think. But it remains true that evolution has, via selection, crafted a specific organization of processes and anatomy by which the heart works. Further, the working of the heart has had causal consequences in the evolution of the biosphere.

Now two points: First, the *organization of the heart* arose largely by natural selection, which, as we will see, cannot be reduced to physics. Thus the *emergence of the organization of the heart* cannot be reduced to physics alone. But it is this very organization of the structure and functions of the heart that allow it to pump blood. Thus, the existence of this organization, the heart in the universe, cannot be reduced to physics. The issue about the inadequacy of reductionism is not whether, given a heart, all its properties can be reduced to chemistry and physics without new, spooky physical properties. Presumably such a reduction can and eventually will be carried out. Rather it is threefold: How did the heart come into existence in the universe in the first place? Second, there is no way the physicist can distinguish that the pumping of blood is the function of the heart. Third, things that have causal consequences in their own right are real. Hearts, by virtue of the organization of structure and processes that they have due to their evolution by natural selection, do have causal consequences as hearts. Hearts are thus real entities. So, too, are organisms and the biosphere. All are part of the furniture of the universe.

But the above argument turned on the proposal that the physicist could not deduce the detailed, specific evolution of this biosphere. If she could, then she could predict the evolution of the heart and ascribe it its function without digging for fossils, which would stand as experimental confirmations of her theory. Can she predict the detailed specific evolution of this biosphere? I will argue that the answer is no in general, and stunningly no in the cases of Darwinian preadaptation, which can be neither prestated nor foreseen and whose detailed specific evolution is not deducible from physics or anything else. Their evolution is partially lawless. Preadaptations are part of the ceaseless creativity of the biosphere. It is because preadaptations cannot even in principle be deduced that both emergence and creativity in the universe are real, and we find ourselves building a new scientific worldview.

Here we enter into more technical material. How could the physicist "deduce" the evolution of the biosphere? One approach would be, following Newton, to write down the equations for the evolution of the biosphere and solve them. This cannot be done. We cannot say ahead of time what novel functionalities will arise in the biosphere. Thus we do not know what variables—lungs, wings, etc.—to put into our equations. The Newtonian scientific framework where we can prestate the variables, the laws among the variables, and the initial and boundary conditions, and then compute the forward behavior of the system, cannot help us predict future states of the biosphere. You may wish to consider this an epistemological problem, i.e., if we only had a sufficiently powerful computer and the right terms to enter into the right equations, we could make such predictions. Later, when we get to Darwinian preadaptations, I will show that the problem is much more than epistemological; it is ontological emergence, partially lawless, and ceaselessly creative. This shall be the heart of the new scientific world-view I wish to discuss.

Without a set of equations about the evolution of the biosphere, the best approach would appear to be to simulate the entire evolution of our very specific, historically contingent biosphere.

The biological world straddles the quantum-classical boundary. For example, about seven photons cause a rod in your retina to respond. And random cosmic rays from outer space, quantum phenomena, cause mutations that may become heritable variations and participate in evolution. Thus any simulation of the evolution of the biosphere must include both earthbound quantum events and cosmic rays arriving from distant, unknown regions of space. It is clear at the outset that the random arrival of cosmic rays from anywhere in spacetime able to causally influence us on earth, that is, our past light cone, is completely unpredictable. In part this unpredictability reflects the quantum character of such cosmic rays, and "throws of the quantum dice." Thus such a simulation is impossible and the physicist cannot even get started on her simulation. However, let us, for the moment, ignore them and focus on the environs of Earth.

Physicists since Einstein have united space and time into spacetime. Virtually all physicists assume that spacetime is what mathematicians call continuous. Continuous spacetime is beyond what is called first-order infinite, an idea I need to take a moment to explain. Georg Cantor, a late-nineteenth-century mathematician, wished to explore the concept of infinity and began by exploring the integers. He showed that if you order all the even numbers—0, 2, 4, etc.—in an infinite sequence, you can pair them with the full set of integers forever. As he put it, the even numbers could be "put into one to one correspondence with the full set of integers out to infinity." On the basis of this correspondence, Cantor said that the even integers are the same "order of infinity" as the integers. One could say, somewhat inaccurately, that there are exactly as many even numbers as integers in the first order, or a denumerably infinite set of integers. Of course, the counting would never stop, because the set of integers is infinite. The same is true for the rational numbers, numbers that can be expressed as a ratio of whole integers, like $3/4$, $4/3$, and so on. By a clever construction, Cantor showed that one could count the first, second, third, and so on, rational number and number them all out to infinity. They, too, can be counted one by one, and each one paired with an integer. Thus the rational numbers are of the same order of infinity as the full, infinite set of integers. All are what Cantor called first-order infinite.

But then there are irrational numbers, which are not expressible as the ratio of two whole numbers—for example, pi, the ratio of the circumference of a circle to its diameter, or the square root of 2. Rational numbers, upon division, yield repeating remainders ($3/4$, for instance, can be written as 0.750000 . . .), but irrational numbers do not; they go on forever without repeating. It turns out that the continuous number line, or real line, as it is called, is dense with both rational and irrational numbers. Roughly, dense means that between any two rational numbers on the real line you can put as many more rational numbers as you wish. The same is true for the irrationals. Cantor brilliantly succeeded in showing that you cannot count the irrational numbers with the integers. Thus, the "order of infinity" of the real line is greater than that of the integers. The real numbers are the second order of infinity. Further, it turns out that if you divide the first-order infinite number of rational numbers by the second-order infinite number of irrational numbers, the ratio is zero. That is, there are infinitely more irrational numbers than the infinite number of rational numbers.

Alan Turing proved years ago that most real numbers could not be computed—that is, he showed that there was no effective procedure, or algorithm, to compute them. Algorithms cannot do everything—they cannot compute most real irrational numbers. This will be of interest when we consider whether the mind is algorithmic. And it bears on the philosophers' idea of supervenience, the infinite list of statements in the lower-level language that can replace the higher-level statement. If most irrational numbers are not computable, I also argue that there is no algorithm to generate the supposed infinite list of statements in a lower level, or more basic, language that will "reduce" a statement in a higher language to the lower one.

Thus the physicist *cannot simulate the evolution of this specific actual biosphere*, even ignoring random quantum disruptions from deep space. Among all the possible throws of the quantum dice there is a second-order infinity of alternative events whose probabilities are achieved by squaring the amplitudes of the appropriate Schrödinger equation, where that equation extends over continuous space. The physicist would have to carry out infinitely many—indeed, a second-order-infinitely many—simulations in order to model our specific biosphere with perfect precision. Obviously, no one could get this much time on a supercomputer. Worse, the simulation still would not have included mutations caused by random cosmic rays from deep space. There is no way to do so. Finally, how could we ever conceivably ascertain which of the infinitely many simulations actually corresponded to the exact sequence of quantum events in this specific biosphere? We cannot. Yet were the quantum events even slightly different, different mutations or other events might arise. In short, given continuous spacetime, there are a second-order infinity of possible histories of the biosphere. Even if we could simulate just one of them, we cannot simulate all of them, and we cannot tell which simulation is the right one.

So it seems the physicist is out of luck. She cannot write down equations and solve for the forward evolution of the biosphere to deduce the occurrence due to natural selection of specific organs such as the heart, and she cannot simulate the evolution of this specific biosphere with its hearts. Such ab initio simulations offer no pathway from fundamental physics to the selective evolutionary emergence and therefore function of the heart. In short, the physicist cannot tell us why hearts exist. Biology is therefore not reducible to physics. Biology is both epistemologically and ontologically emergent.

Again, this does not mean that any physical laws are violated. Why should we care? Because the evolution of a specific biosphere, say ours, takes us firmly beyond reductionism. Hearts exist in the biosphere, neither their occurrence nor their function can be accounted for by physics alone, yet they have real causal consequences in the further evolution of the biosphere, including in our own lives. Hearts are really part of the universe. So, too, are wings, eyes, the opposable human thumb, the tools we make with our hands and minds, and the economy that results. All are epistemologically and ontologically emergent.

Further confirmation that biology cannot be reduced to physics alone is the status of natural selection. Natural selection rests on reproduction with heritable variation in a context of constrained resources. Darwin knew almost nothing about reproductive physiology or the basis of heritable variations; thus he could not use such ideas as DNA or genetic mutations to build his theory. Now we know considerably more. Many of us are even involved in trying to create life anew. The efforts are wonderful. We are likely to succeed in the next hundred years in creating not only molecular reproduction but life itself. Already self-reproducing DNA and protein systems have been created.

The central implication of these efforts is that life is almost certainly not constrained to current terrestrial life. Indeed, current life, with DNA, RNA, proteins, transcription of DNA into RNA and translation of RNA into proteins, and so forth, is almost certainly too complex to have been the earliest life form to appear on Earth or elsewhere. Life must have evolved from simpler beginnings. Thus, life is almost certainly broader than current terrestrial life—and a general biology of life in the cosmos awaits our discovery. NASA's program in astrobiology, begun about a decade ago, represents one of the first steps towards such a general biology. Astrobiology seeks, among other things, characteristic signs of life that we can look for in our explorations of space. In chapter 8 I will discuss laws of self-organization that suggest that contemporary cells have evolved by natural selection to a highly improbable but advantageous state called dynamically critical. It is a candidate law that "criticality" may be a feature of life anywhere in the cosmos. It would be a triumph to find universal laws of organization for life, ecosystems, and biospheres. The candidate criticality law is emergent and not reducible to physics alone. Other such laws, if we find them, would presumably also not be reducible to physics alone.

Returning to Darwin's natural selection: if life can be highly variable across the cosmos, this means that natural selection transcends any specific physical realization of it. Darwin is completely agnostic about the physical basis of heritable variation. Once again we find ourselves facing a multiple-platform argument. Since natural selection can "run" on many physical realizations of life, it cannot be reduced to any one of them. More, we have no idea of the diversity of physical systems that can be alive and subject to natural selection. Thus natural selection cannot be reduced to any necessary and sufficient set of statements about this or that set of atoms or molecules. Life may still be possible even if the constants of nature change a bit. If so, natural selection of living forms transcends this specific universe, hence cannot be reduced to it.

Again, I am not claiming that natural selection violates any law of physics, merely that it cannot be reduced to physics. While any one case of life and evolution is a specific realization, Darwin's idea is not confined to that realization. It stands on its own as a statistical "law" that applies wherever there are restricted resources and descent with heritable variations. It does not need physics to be understood. Indeed, Darwin used no appeal to physics to state his idea of natural selection. It can even be stated mathematically, although it's well known that Darwin put no equations in *The Origin of Species*. His law does not need physics, except in the sense that real organisms are built of some physical stuff. It is a law of biology, not of physics, and not reducible to physics alone.

As I mentioned earlier, philosophers have advanced a weakened form of reductionism, supervenience, in which in the reducing theory an *infinite*, not finite, list of statements at the lower level is necessary and sufficient for the statement at the higher-level theory. If supervenience works as a reduction procedure, then the statement "hearts evolved by natural selection" is a shorthand statement at the higher level language which is to be replaceable by the infinite list of statements at the lower, physics level. But as I have already argued, supervenience is not helpful in an actual reduction from a higher-level description to a lower-level description. We have a recipe for writing down the infinite set of integers, 1, 2, 3 . . . infinity, even if we cannot complete it. For the second-order list of all possible irrational numbers, we do not even have a recipe. That is Cantor's proof and Turing's proof that most irrational numbers are not computable. We also have no recipe for writing down the possibly infinite or second-order-infinite list of all possible organisms to which natural selection would apply in this and perhaps similar universes. So it seems utterly useless to appeal to such an infinite supervenience list to "reduce" Darwin's law.

There is another way to see the independence of Darwin's law from underlying physics. It is sometimes supposed that science proceeds by induction. Scientists see one white goose and then many white geese and hypothesize that all geese are white—the step to "all geese are white" being the inductive step. Now suppose a superphysicist could deduce from string theory or the standard model to a tiger chasing a gazelle, and also deduce the heritable variation in the paws of the tiger's offspring, and that one such variant would be fitter than the rest. How would the physicist reason by induction from the specific case of the tiger to Darwin's general law about heritable variation and natural selection? Presumably he would find many cases of natural selection and induce that it is a general "law." Now we are quite convinced that natural selection would apply to any organism in the universe. Darwin himself did not reason as does our hypothetical inductive physicist. Darwin read Malthus's argument that the rate of growth of food production is only linear while the growth of populations is geometric, and postulated that we must eventually outrun our food supplies and suffer famine. From this, Darwin jumped to envision a general competition within each species for finite food resources, and then, on the supposition of heritable variations, he *inferred* his law of natural selection. What he did *not* do was to "inductively" examine many cases in the wild, find the heritable variations, show that they were selected by the competition he envisioned, and then arrive inductively at the conclusion that natural selection applies to all life.

How then would the superphysicist get from his deduction of the tiger chasing the gazelle to Darwin's general law? He has not read Malthus, we may suppose, but he will proceed by induction, examining many individual cases and then guessing that some frequently observed phenomenon is general or universal. If so, the physicist has an incredible number of deductions to perform in order to get from string theory to tigers, gazelles, turtles, palm trees, and so on, before he can begin to make the inductive generalization from specific cases of natural selection to Darwin's laws. And, even then, he cannot generalize to neighboring unknowable universes. He cannot arrive at Darwin's law, in all its generality across similar universes, by induction alone. Alternatively, the physicist can go through the reasoning that Darwin says he did, and *reinvent natural selection at its own level, the level of organisms* in selective environments. But then natural selection is not reduced to the infinite list of possible organisms in possible universes, nor is it in any way useful to try by supervenience to reduce it to those physical laws. Darwin's law stands in its own right at its own level—biology.

To state the matter even more provocatively: you cannot get to Darwin from Weinberg. Weinberg rests his reductionism on the claim that the explanatory arrows always point downward, from society to cells to chemistry to physics. But with respect to evolution of hearts by natural selection, do the explanatory arrows actually point downward to string theory or whatever is down there? No. They point upward to the *selective historical* conditions in the actual evolution of organisms in the specific biosphere that gave rise to hearts. So Weinberg's explanatory reductionism is not sufficient. With Anderson, Laughlin, and Kadanoff, I want to say that any putative reduction of biology to physics is not needed and cannot be carried out. We have now moved beyond reductionism and arrived at emergence, both epistemological and ontological.

These limitations on reductionism are terribly important, for they imply that emergence is real. Biology is really not just physics. Nor are organisms nothing but physics. Organisms are parts of the furniture of the universe, with causal powers of their own, that change the actual physical evolution of the universe. Biology is emergent with respect to physics. Life, agency, value, meaning, and consciousness have all emerged in the evolution of the biosphere, as I will discuss later.

That reductionism is limited, however, does not mean it is not powerful, amazingly productive, and tremendously useful scientifically. We simply need to understand its place, and recognize that we live in a very different universe from that painted by reductionism alone.

5

THE ORIGIN OF LIFE

God's hand, finger outstretched, reaches towards Adam's finger on the ceiling of the Sistine Chapel, an image of the Creator passing the breath of life to humanity. Central to most creation myths is the view that life is a gift from God or the gods. The Pueblo speak of sipapus—sacred holes in the land that stretches from Mexico City to north of Santa Fe. According to the legend, the first people passed through sipapus from the underworld and populated this world. Interestingly, the Spaniards, after conquering the Pueblo, erected the famous Santuario de Chimayó directly over the northernmost sipapu, where a sacred hole is said to be the source of holy sand. I have filtered this soft earth through my fingers in the Santuario, amid abandoned crutches hanging on the walls. We build our churches on the holy sites of other cultures and install our own gods. Notre Dame sits on a holy Druid site. "Ours is the true God," we say by doing so. But this usurpation is also a kind of acknowledgment, a perverse honoring of the more ancient culture: a holy site is a holy site, just as a holy day is a holy day. In the same way, our scientific view of the origins of life overlies ancient creation myths. That is why I feel we must use the God word, for my hope is to honorably steal its aura to authorize the sacredness of the creativity in nature. May we find the creativity in nature sacred whether we are atheists or believers in a God who breathed life into this universe of ceaseless creativity.

Almost all scientists are persuaded that life is a phenomenon that arose naturally in the universe, probably in myriad places, and that it almost certainly arose spontaneously on Earth some 3.8 billion years ago, soon after the Earth's crust cooled sufficiently to support liquid water. If life is natural, as I firmly believe, then part of the immense call for a transcendent Creator God loses its force. If we seek a reinvented sacred based on this universe and its miraculous creativity, then a natural explanation for the origin of life in this universe is of paramount importance. More, the science is exciting. If we succeed in creating life anew or find it somewhere else, the way will be opened to a general biology concerned with life anywhere in the cosmos, freed of the constraints of terrestrial biology. Such a general biology will stand astride physics, chemistry, mathematics, and biology and change all of these sciences. Who knows what marvelous general principles may be found? I have already mentioned one candidate law, the dynamical criticality that may characterize modern cells, which I will discuss in chapter 8. Others are discussed in my book *Investigations*, which suggests four candidate laws for any biosphere: the criticality of cells; the possible self-organized criticality of the biosphere and our economy (which I'll discuss in this book in chapter 11); the organization of ecosystems; and the possibility that biospheres may evolve to maximize the diversity of organized events that can happen. We will see early hints in chapter 7 of this last candidate law concerning mature ecosystems, where total energy flow times a mathematical measure of the diversity of energy flow in the ecosystem, appears to produce a high value. As I will discuss in chapter 7, where I will also describe the physicist's concept of work, just possibly, a similar "law" may arise with respect to something like total work times the diversity of work that evolved cells can carry out. Thus, if the diversity of life tends to increase despite extinction events, as it has actually done on Earth, a similar law could apply to entire biospheres, maximizing the diversity of what can happen. These may be laws that apply to any biosphere in the universe. Who knows what marvels of life-forms we may find? We may well find that we can make life, capable of coevolution, and yet cannot predict what will happen! What more stunning experimental confirmation might we hope for that the biosphere is unspeakably creative?

Work on the origin of life has been underway for a half century, substantial progress has been made, and it is likely that we will create life anew sometime in the next hundred years. But while these two efforts are obviously related, it's important to point out that that relation is more tenuous than people often think. Creating life experimentally might give us the keys to the origins of life on Earth, but it might not.

First, if we create life experimentally, it does not follow that we will have established the actual historical route to life on Earth. We may never know that history. The fossil record of organisms and ancient molecular species may be insufficient to tell us. Second, what we create as simple life, for example, self-reproducing systems of peptides, proteins, or simple organic molecules, will be far simpler than current life. The self-reproducing molecular systems of proteins we have been able to create in experiments lack the familiar process of protein synthesis, via transcription of DNA to RNA and translation of RNA to proteins, the central dogma of molecular biology. In living cells, rather amazingly, among the proteins encoded by genes are those very proteins that carry out translation, including translation of themselves. So it takes DNA, RNA, and encoded proteins to carry out the very process of translating genes into proteins. In short, in contemporary cells, the molecular mechanisms by which cells reproduce form a complex self-referential system that clearly is highly evolved.

If we create much simpler life, this will show that life can have manifold physical realizations. We are back to the multiple-platforms argument: life is not reducible to any specific physics. Already, as we will see below, more than one class of chemical compounds, DNA and small proteins, can each form self-reproducing systems. Thus, it is already clear that molecular reproduction can run on multiple platforms. While these examples are not yet "life," it is virtually certain that life, too, can run on multiple physical platforms. Then life is not reducible to any specific physics, but is ontologically emergent. Equally wonderfully, a general biology for life anywhere in the cosmos beckons.

Contemporary cells are what are called nonequilibrium physical chemical systems. The idea of chemical equilibrium is familiar to all of us who remember two chemicals, say X and Y, that can interconvert. At chemical equilibrium, the rate at which X changes to Y just equals the rate at which Y changes to X, so there is no net change in the relative concentrations of X and Y except for minor fluctuations which damp out. Cells maintain chemical concentrations that are far from the equilibrium point. To do this they take in matter and energy, hence are what are called open thermodynamic systems.

In addition, cells do thermodynamic work cycles, much like a steam engine that converts a temperature difference, or gradient, into mechanical motion. Cells have a complex metabolism linking the synthesis and transformation of thousands of small organic molecules in complex interconnected pathways. The individual chemical reactions are themselves speeded up by enzyme catalysts. The origin of such a connected metabolism needs to be accounted for.

Most notably, cells can reproduce themselves. At least part of this is due to the well-known workings of the central dogma of molecular biology, noted above, in which DNA structural genes are transcribed into RNA molecules that are then translated into proteins via the genetic code. But the self-reproduction of a cell is vastly more complex than the central dogma and involves what I will call a propagating organization of linked processes. These processes complete a set of work-related construction tasks that closes on itself such that the cell as an entire entity reproduces itself. In so doing, the cell replicates its DNA using protein enzymes, synthesizes its bounding membrane, and re-creates a variety of intracellular organelles such as mitochondria, the major site of energy generation. I will discuss this puzzling "propagating organization" of process in chapter 7. It is a subject today's science hardly knows how to talk about. This chapter, however, focuses on a different issue: molecular self-reproduction in very simple systems that are promising starts to understanding the whole story.

The modern era of origin-of-life research started with remarkable experiments by Stanley Miller. Miller was trying to synthesize, in conditions resembling the early Earth, the diversity of small organic molecules that are the building blocks of the more complex molecules of life, such as the amino acids that link in linear chains to form proteins. He took a retort with a mixture of gases and water thought to mimic the primordial atmosphere and introduced energy in the form of electric sparks. Wonderfully, in a few days, a brown sludge formed in the retort. Analysis of this material showed an abundance of several amino acids found in current life. Miller's seminal experiment opened a continuing quest to understand the prebiotic synthesis of a variety of small organic molecules found in life, such as amino acids; sugars; the nucleotides that form DNA and RNA; the lipids that form cell membranes; and so on. To summarize decades of work, most of the small organic molecules of life can in fact be synthesized under reasonable prebiotic conditions, although a certain magic remains hidden from view. The conditions that lead to one class of molecules may not lead to other classes of molecules, leaving open the question of how all these came together.

Some of the organic building blocks may have come from space. Astronomers have discovered giant, cold molecular clouds of unknown chemical diversity that are the birthplaces of stars. Large atoms (i.e., everything heavier than lithium) are formed inside stars and then released into space when the stars eventually explode as supernovae. These atoms seem to form complex molecules in the giant molecular clouds. Organic molecules have been found abundantly in meteorites that have fallen to the surface of the Earth. It may be that myriad small organic molecules, and even more-complex molecules, fell onto the young Earth and supplied some of the early molecules of life. For example, carbonaceous meteorites have complex carbon-based molecules and even contain lipids that can spontaneously form bilayered, hollow vesicles called liposomes when immersed in water. These liposomes are astonishingly similar in structure to the cell membranes surrounding contemporary cells.

The generation of a set of organic molecules that are constituents of life is a wonderful first step. But perhaps the first central question about the origin of life is the onset of molecular reproduction. When, and how, did molecules appear that were able to make copies of themselves, either singly or as interacting systems of molecules?

The first idea concerning molecular reproduction was that life must be based on template replication of DNA, RNA, or some similar molecule. The structure of these famous molecules was, as we know, first described by James Watson and Francis Crick in 1953, when they showed that DNA is the well-known double helix. It is comprised of four types of building blocks, called nucleotides or bases, abbreviated as A, T, C, and G. The double helix consists of two "ladder sides," each made of linked sugar phosphate parts and joined by specific pairings of A to T and C to G. These pairs form the rungs of the ladder.

The DNA double helix has two amazing features. First, it is essentially perfect in that the A-T and C-G pairs are each of the same length, leaving the two ladder sides the same distance apart whatever the nucleotide sequence is. Second, this perfection allows the linear sequence of A, C, G, and T along a single side of the double-stranded helix to be arbitrary. Any linear arrangement of the four bases can, in principle, occur. A gene is a sequence of bases.

During the transcription process, when the cell is carrying out its normal functions, the sequence of nucleotides in the DNA is "copied" into a cousin molecule, RNA, forming a single strand, with the minor alteration that a U substitutes for the T in DNA. In the next step, sets of three adjacent RNA nucleotides constitute genetic "codons," each of which specifies one from among twenty types of amino acids. These are built into a long chain of amino acids that constitutes a protein. The protein later folds into a three-dimensional structure and is then able to carry out its function in the cell.

In normal cell reproduction, the double-stranded DNA is unwound into two single strands, each with a sequence of A, C, T, and G along its length. Each of the two single strands of DNA is then replicated by an enzyme called a DNA polymerase, creating a second strand, matching each A on the first strand with a T on the second strand, each C with a G, and so on. The actual process is very complex, but this is its essence.

With the discovery of DNA's structure, many scientists were so struck by the beauty and symmetry of this molecule and its elegant mode of replication that they felt all life must be based on double-helix template replication. This conviction has driven the longest line of experimental investigation in the field. Leslie Orgel, an outstanding organic chemist, and many other chemists have sought to use single stranded RNA molecules and induce them to reproduce by template replication in the absence of a protein enzyme. Here the idea has been to start with some arbitrary single-stranded sequence of nucleotides along an RNA molecule and supply free nucleotides in the reaction mixture. Then the hopes are that (1) the sequence of bases along the template RNA molecule, say A, A, C, G, G, U, G, C . . . ,would successively line up the matching free nucleotides, U matching to A, C matching to G, and so on; (2) this matching would lead to the formation of proper bonds between successive nucleotides to create a new, complementary second strand; (3) once formed, the two strands would separate; (4) once separated, each strand would repeat the steps above to create a population of replicating RNA molecules. The point of the "arbitrariness" of nucleotide sequences along the RNA template is that replication must be independent of the sequence, just as the process that printed this book is independent of the words on each page. The arbitrariness of the sequence of nucleotides is what allows DNA to carry "selected genetic information."

This line of investigation looks like it should work. But after forty years of effort, applying all sorts of tricks, its proponents have not yet succeeded. They may yet, but hope is dwindling. No one has gotten more than a modest number of nucleotides to replicate on the new strand in the absence of a protein enzyme like RNA polymerase. This failure is a sad thing. The line of thought is so straight, the symmetry of DNA and RNA so beautiful, that the strategy should have worked. Some of the chemical pitfalls are now well understood, but non-enzyme-catalyzed direct-template replication of DNA or RNA looks far less promising than it once did.

A second line of work, one that has captured the attention of many chemists and biologists, is based on what is called the RNA world. In this view, early life consisted entirely of RNA molecules, or RNA and a simple metabolism to support the synthesis of RNA from organic precursors. This idea grew out of an unexpected discovery. It is widely known that proteins commonly act as enzymes, speeding chemical reactions, and that RNA molecules can carry genetic information. But it was not expected that RNA molecules could act as catalysts. Yet they can. Such molecules are called ribozymes. How they were discovered is a fine story that ends with Nobel prizes, but it is not our current concern. The cause of excitement was that the same class of molecules, single-stranded RNA molecules, could both carry genetic information and catalyze reactions, including reactions among RNA molecules themselves. And it turns out that current cells across all of the kingdoms of life show what are thought to be RNA "fossils," molecular remnants of a time when life was based almost exclusively on RNA. I say "almost" because these early cells would have also required a metabolism to form nucleotides, as well as energy sources.

An early statement of the RNA world view was made by Nobel laureate Walter Gilbert, who proposed that, somehow, *systems of RNA molecules could catalyze one another's formation.* As we will discuss below, my own theory of the emergence of collectively autocatalytic sets is one potential means by which such a system of mutually catalytic ribozymes might form.

After the RNA world theory became popular in the origin-of-life field, in the 1990s, researchers switched attention from the idea of a system of RNA molecules that could catalyze one another's formation and instead began trying to evolve single RNA molecules able to replicate arbitrary RNA single-stranded sequences. Such a *single* RNA molecular sequence would be an *RNA polymerase able to copy any RNA molecule, hence able to copy itself and reproduce.* The procedure to evolve such an RNA polymerase is exciting—a kind of in vitro evolution. One creates a vast diversity of single-stranded RNA molecules, called a "library," and uses clever techniques to select out of this mixture those rare RNA molecules that might catalyze the desired reaction. David Bartel, a molecular biologist, is using this procedure to seek out, from among 10^{15} RNA molecules, one that can add a sequence of free nucleotides to an arbitrary single-stranded RNA template. He has partially succeeded, finding an RNA molecule that can catalyze the sequential addition of about thirteen nucleotides to a specific RNA template. Bartel's work creates the real hope that an RNA polymerase can be found that can copy arbitrary RNA sequences, including, therefore, single-stranded copies of itself. This would be a single self-reproducing RNA sequence.

Future experiments will explore this possibility. But I have a concern: catalysis is noisy. Consider an RNA polymerase. At each step in copying an arbitrary RNA sequence, the correct one nucleotide among four

must be added to the growing new chain. As Jerry Joyce and Leslie
Orgel were the first to point out, this is not easy, for at any step the
wrong nucleotide may be added. Suppose the RNA polymerase made
copies of itself with mistakes. That would tend to make the new molecules even sloppier ribozymes, generating yet more errors in the next
generation's copies. These copies would generate a still-higher error
rate, until eventually the daughter sequences would completely lose the
ability to function as polymerases. This could lead to what Orgel called
an error catastrophe, in which errors in the population of RNA ribozyme polymerases would accumulate so fast that natural selection
would be overwhelmed and the RNA polymerase function lost in a sea
of useless sequences. It has not yet been assessed whether this is a reasonable worry. Perhaps some threshold of reproduction precision exists
above which the RNA ribozyme polymerase population can sustain itself, below which an error catastrophe sets in. No one knows as yet.

I have a second concern. The probability of such an RNA polymerase
appearing seems to be less than 1 in 10^{15}. Thus it is a rare molecule indeed.
Given a sufficiently large diversity of RNA sequences, one might arise by
chance and then take over the population by replicating itself—or it might
waste its time copying other, nonreplicating RNA molecules nearby and
thus fail to reproduce itself. In short, the conditions for accumulation of
such an RNA polymerase ribozyme have yet to be understood. This does
not mean, however, that these concerns cannot be overcome.

A third issue concerns the initial formation of a single self-reproducing
RNA and the formation of a connected metabolism leading from simple organic molecules to the nucleotides that are the constituents of RNA. The
first problem is not so easy, for thermodynamics does not favor the synthesis
of the proper chemical bonds between free nucleotides and the hoped for
chain of linked nucleotides, called a 3'–5' phosphodiester bond. Instead a
different bond, called 2'–5', tends to be formed that does not lead to the
proper single stranded RNA molecule. Next, even were such a ribozyme
polymerase formed, where would the ribozymes come from that would catalyze the connected metabolism leading to the synthesis of the nucleotides
that form the ribozyme polymerase and the "metabolic ribozymes" themselves? While the answers remain unknown, the standard RNA world view

gives us at least a rough picture of the onset of a connected metabolism that might lead to the synthesis of RNA nucleotides and the origin of ribozymes that can catalyze connected pathways of reactions among metabolites from available building blocks, including sources of chemical energy. Some evidence supports this possibility. Using libraries of random RNA sequences, a number of workers have evolved in vitro RNA ribozymes that can catalyze a surprising variety of chemical reactions. Thus we can imagine a ribozyme polymerase, able to copy itself and all other single stranded RNA sequences, existing alongside ribozymes for metabolism. The RNA world also needs to confine the reactants so that they can interact at high rates, perhaps by synthesis of lipids that can form bounding membranes confining the ribozymes.

Current cells use protein enzymes almost exclusively, not RNA ribozymes. This is probably because proteins are more chemically diverse than RNA molecules, hence far more likely to be able to catalyze any particular reaction. This presumptive virtuosity of proteins does not rule out an RNA world, but it is directly testable. If true, it would tend to support a role for small proteins in early life. The way to test this question is to generate high-diversity sets, or "libraries," of RNA sequences, and random amino-acid sequences that constitute proteins, or shorter versions called polypeptides, or even shorter versions called peptides. We can then test whether it is more common to find RNA or peptide sequences able to catalyze each of a diversity of reactions. An important comment is in order: RNA sequences fold readily. Folding may be necessary for catalytic and other functionalities. In evolved proteins, linear sequences of amino acids fold into three-dimensional shapes, but it was thought that less-evolved proteins did not, at least not easily. Two bodies of work, one by Thomas LaBean, the other by Luigi Luisi, have shown that, in fact, random peptides fold reasonably well. How well is not yet known. My personal bet is that small peptides, particularly those bound to metal ions, called organometallic molecules, will turn out to be quite capable of catalyzing a diversity of reactions. I stress again that this question is testable now, with current techniques.

My hunch is that peptides, including organometallic ones, are far more likely to serve as catalysts than RNA molecules. The reason I think so is related to what are called catalytic antibodies. An antibody is a complex protein made in our immune systems that binds to what is called an antigen. For example, invading bacteria and viruses have molecular features that are recognized and bound by antibodies to fight off infection. Richard Lerner and others reasoned that if an antibody can bind an antigen, perhaps it could catalyze a reaction. Such catalytic antibodies have been shown in experiments. It now appears that the probability that any one antibody molecule, chosen at random from the 100 million different antibody molecules a human can make, catalyzes a given reaction may be about one in a million. This figure, far higher than the apparent probability that a random RNA molecule will catalyze a given reaction, is one reason I think peptides played a fundamental role in the origin of life. Another reason is that amino acids form readily under prebiotic conditions, while DNA and RNA nucleotides do not.

A third body of work yielding reproducing molecular systems, carried out by Luisi, concerns the bilayered lipid membrane vesicles described above, or simpler versions called micelles. In micelles, lipid molecules form aggregates that may have an oily, water-hating (or hydrophobic) core comprised of the lipid molecules' "tails," while their water-loving (or hydrophilic) heads remain outside, exposed to an aqueous environment. Here, there are two driving ideas. The first is that life may have started as self-reproducing micelles or bilayered liposome-like vesicles. The bilayer consists of two lipid layers, where the water-hating, or hydrophobic, tails of the lipids in each layer nestle together. Liposome bilipid membranes are almost identical to cell membranes. The second idea is that any system of complex organic reactions, such as a protometabolism or the collectively autocatalytic sets that I will describe below, must be confined to a small volume if the molecules are to be able to interact at a high rate. A bounding membrane is one obvious way to satisfy this requirement. (Reactions confined to surfaces is another way, for instance, or reactions in small, hollow regions of porous rocks.) A number of workers, notably Luisi, have succeeded in creating liposomes, in the presence of lipid molecules in the aqueous environment, that actually grow in volume and spontaneously divide. They have thus demonstrated molecular reproduction of a simple structure, a model of a cell membrane. It is highly reasonable to expect that this reproduction was part of the early history of life.

A fourth body of work derives from two sets of exciting experiments that have actually achieved molecular reproduction of DNA and polypeptide polymers via what I will call autocatalysis and collective autocatalysis. While the symmetry of the double helix is lovely, I have always believed that the basis of life is deeper and that it rests in some way on catalysis, the speeding up of chemical reactions by enzymes. My second intuition is that life is based on some form of *collective autocatalysis*, in which the molecules in a set catalyze one another's formation. This stands in contrast to the theory that a *single* molecular species—such as the hoped-for RNA ribozyme polymerase that's able to copy itself—is the basis of life.

Twenty years ago, Gunter von Kiedrowski achieved the first example of chemical molecular replication. He used a hexamer, which is a single-stranded DNA sequence with six nucleotides. He designed the hexamer so that the first three nucleotides might bind with a single-stranded DNA trimer—a molecule with three nucleotides. In turn, the second three nucleotides would bind to a second trimer. Then, von Kiedrowski hoped, the fact that the DNA hexamer had lined up and held in juxtaposition the two trimers might catalyze the formation of a bond linking the two trimers into a hexamer. And if, by clever design, it turned out that the second hexamer was a copy of the first, the single strand of DNA would have achieved self-reproduction.

It worked. Von Kiedrowski created the first reproducing molecular system. His hexamer is autocatalytic: it builds a second copy of itself by "ligating" the two trimers into a new hexamer. Unlike Orgel with his one-at-a-time template replication of arbitrary RNA sequences, von Kiedrowski used a specific single-stranded DNA hexamer as a simple enzyme to bind together two single-stranded DNA trimers. While Orgel hoped to get arbitrary sequences of bases to reproduce, on the grounds that such arbitrary sequences constitute genetic "information" (a vexed term we will return to), von Kiederowski is less of a purist. He's less concerned, for now, with "information" than with simply getting some DNA sequence to replicate itself.

But von Kiedrowski accomplished something I consider even more wonderful. He constructed two different single-stranded hexamers, say A and B, such that A catalyzed the formation of B from two trimer fragments of B, while B catalyzed the formation of A from two trimer fragments of A.

This step is crucial. Let us call this AB system collectively autocatalytic. It is essential that no molecule in the AB system catalyzes its own formation. Rather, *the system as a whole*, A, B, and the fragments of A and B, are jointly autocatalytic. A kind of *catalytic "closure"* has been achieved, in which each molecule has its formation catalyzed by some molecule in the system.

If A and B can form a collectively autocatalytic set of two molecular species (plus their building blocks), why cannot three, ten, or ten thousand molecules form a collectively autocatalytic set? They can. A cell in your own body is just such a collectively autocatalytic set of thousands of molecular species. No molecule in it catalyzes its own formation. The cell as a whole collectively catalyzes its reproduction.

Von Kiedrowski's results are exciting, but still based on "template" recognition of DNA hexamers and trimers. The idea that proteins might be capable of molecular reproduction had seemed remote, since the folded three-dimensional form of a folded protein seemed to offer no way for reproduction to occur. Nevertheless, molecular reproduction has now been shown in experiments to be possible, based on collective autocatalysis among small proteins. Template replication is not necessary for molecular reproduction. This means that the diversity of molecular systems capable of molecular reproduction is almost certainly far greater than has been supposed. So, too, are the chances for life in the universe.

Proteins do indeed seem poor candidates for molecular reproduction. A protein's "primary sequence" is a linear sequence of some twenty kinds of amino acids, typically folded up into a three-dimensional shape. But proteins have no obvious analogue to DNA's A-T and C-G base pairing to guide anything like template replication.

Proteins, when they fold, can form a variety of structures. One common structure is called an alpha helix. The chemist Reza Ghadiri used a specific thirty-two-amino-acid helix that had the added property that it folded back on itself to form a "coiled coil." He hoped that since the thirty-two unit sequence could bind to itself, it might bind to two smaller fragments of itself, lengths of fifteen and seventeen amino acids. Then, by holding these fragments next to one another, he hoped that the thirty-two-amino-acid sequence might catalyze the formation of a proper peptide bond between the two fragments to produce a second copy of the initial thirty-two-unit sequence.

It worked! The results are truly historical, for they demonstrate unambiguously that molecular reproduction does not need to be based on the beautiful symmetry of DNA or RNA, or on arbitrary template replication in the Orgel fashion. Life may be vastly wider in the universe. Again we come to a multiple-platform argument: it appears that life, like computation, may be independent of the underlying physics.

Like von Kiedrowski, Ghadiri has gone further. He has succeeded in creating collectively autocatalytic sets of peptides, in which peptide A catalyzes the formation of peptide B from B fragments, while B catalyzes the formation of A from A fragments. The fragments are supplied by Ghadiri to the reaction system. Again, catalytic closure has been achieved. For every molecule in the set whose formation requires catalysis, there is a molecule in the set that carries it out. This is important for at least two reasons. Contemporary cells, as I have stressed above, are collectively autocatalytic wholes in which no molecular species catalyzes its own formation. Experimental demonstration of collective autocatalysis by small proteins thus shows us that it is catalytic closure of a set of molecules that is the backbone of life. Below I will describe a theory, now fully open to experimental test, that says the emergence of such collectively autocatalytic polymer systems is to be expected. If correct, this theory suggests that molecular reproduction may be an emergent property of complex chemical-reaction networks, hence life may be far more probable than we have thought. In turn, this emergence is not reducible to physics alone.

So Ghadiri has once and for all demonstrated collective autocatalysis in peptides. His results open the way for us to consider that many different sets of molecules, ranging from small organic molecules to organometallic molecules, to peptides, to RNA and DNA, might form collectively autocatalytic sets and be the basis of life on Earth and elsewhere in the cosmos. It is deeply important that these collectively autocatalytic sets are examples both of *catalytic closure,* a property that resides in no single molecule but in the set as a whole, and also of *kinetic control.* Chemists distinguish between chemical process that are or are not under kinetic control. Reactions without kinetic control are generally those that are not catalyzed and, in addition, are very near chemical equilibrium. The rate of the reaction depends on the concentrations of substrates and products. Reactions under kinetic control are those that are catalyzed in a particular direction, for instance, by an enzyme, so that the abundances of the molecular species are quite displaced from equilibrium. The reaction in one direction can be speeded up or slowed down substantially.

Kinetic control in collectively autocatalytic sets means something of very general importance. Collective autocatalysis with kinetic control is an example where the *integrated system constrains the kinetic behaviors of its parts and organizes the kinetic behaviors of the chemicals of which it is made. These constraints yielding organization of process are partially causal in what occurs.* Thus these collectively autocatalytic systems are very simple examples of the kinetic organization of process, in which what might be called *the causal topology of the total system constrains and guides the behavior of its chemical constituents.* These constraints, imposed by the system's causal topology on the kinetics of its parts, are a case of *"downward causation."* Because these constraints are partially causal, the explanatory arrows do not point exclusively downward to the parts but also *point upwards to the organization of the whole.* The whole acts on the parts as much as the parts act on the whole. Collectively autocatalytic systems are perhaps the simplest example of philosopher Immanuel Kant's idea that in an organized being, the whole exists for and by means of the parts, and the parts exist for the whole. Kant was speaking of organisms. So am I. An evolved cell is an autocatalytic whole, whose organization of processes and closure of causal tasks to reproduce came into existence via heritable variation and natural selection. This evolutionary process, we have seen, cannot be reduced to physics alone. Thus, the emergence in the universe of collectively autocatalytic, evolved cells and their "topology" of organization of kinetic-controlled process is ontologically emergent, and the same topology of kinetic control of the "whole" is partially causal in constraining the kinetic behavior of the parts. The explanatory arrows do truly point upward, and an evolved cell fulfills Kant's dictum.

My own theory of collectively autocatalytic sets suggests that their formation is highly probable. The theory can now be tested. If correct, the routes to molecular reproduction may be much easier than we have imagined, and constitute a form of fully emergent, spontaneous self-organization of a chemical-reaction system. Such emergence would not be reducible to physics. And life, in the sense of molecular reproduction, would be expected, not incredibly improbable. If so, our view of life changes radically. Not only does life not need special intervention by a Creator God, it is a natural, emergent expression of the routine creativity of the universe.

THE THEORY OF COLLECTIVELY AUTOCATALYTIC SETS

I will describe my theory of the spontaneous formation of collectively autocatalytic sets two times: first in its initial form, then in a broadened form. The reasons for discussing it are threefold.

First, it may be right. This would indicate that achieving molecular reproduction is easier than we have thought, even highly probable, given a sufficiently diverse chemical-reaction system. Since the theory is testable, we may soon know if it is correct for complex chemical-reaction networks, and it may yield insights into the probability and routes to the origin of life on Earth and in the cosmos.

Second, the mathematics behind the theory is not reducible to physics. The theory is organizational, and, if true, a candidate organizational law that could apply to life universally. It rests on mathematics and general properties of molecules, and not on the specific molecular embodiment. As I mentioned above, the multiple-platform argument comes in, for the theory implies that many different molecular species can achieve collective autocatalysis. Therefore, it cannot be reduced to any specific underlying physics.

Third, the theory is our first example of fully emergent, spontaneous self-organization and its potential role, with natural selection, in evolution. Since Darwin's time almost all biologists have felt that selection is the sole source of order in biology. But a growing number of scholars are suggesting that even the abiotic world exhibits astonishing self-organization. Snowflakes, with their evanescent six-fold symmetry, are an example. The theory of autocatalytic sets is an example at the dawn of life itself: complex polymer systems may spontaneously organize into collectively autocatalytic self-reproducing systems. Self-organization may require that we rethink all of evolutionary theory, for the order seen in evolution may not be the sole result of natural selection but of some new marriage of contingency, selection, and self-organization. New biological laws may hide in this union.

At its base, the theory of the emergence of collectively autocatalytic sets rests on several ideas: (1) a set of molecules, (2) a set of possible reactions among the molecules, (3) the ratio of reactions among the molecules to the number of molecular species, and (4) within the set of molecules, the distribution of which molecules catalyze which reactions. With this information in hand, we can ask whether a collectively autocatalytic set exists within the set of molecules. In particular, as the diversity of molecules in the system increases, what is the probability that the system has an autocatalytic set?

Under very general assumptions, the theory shows that as the diversity of molecules in the system increases, the ratio of reactions among the molecules to the molecules increases, and therefore the number of catalysts in the system for the reactions among the molecules gradually rises above a threshold at which the likelihood of collectively autocatalytic sets in the system becomes nearly certain. In short, at a certain point, that is, when the diversity of molecules is high enough and the ratio of reactions to molecules is high enough, it becomes expected that each molecule has at least one reaction leading to its formation catalyzed by at least one member of the reaction system. At this point the emergence of collectively autocatalytic systems becomes a near certainty. The details of the complexity at which collectively autocatalytic sets are expected to emerge depend on the details of the distribution of catalytic capacities among the molecules in the reaction system. This is a powerful case of self-organization that is not reducible to physics, and an organizational law of the Laughlin type. It offers an account in which the origin of life is natural, emergent, and expected. I will focus first on peptides and RNA, the forms of molecules in the initial theory, but in fact the theory can apply to reactions among a wide variety of molecular species. While I favor peptides as the candidate molecules, the RNA verrsion of the theory is compatible with Walter Gilbert's early statement of the RNA world in which molecules within a system of RNA molecules form one another.

The mathematicians Paul Erdos and Alfréd Rényi in 1959 developed the theory of random graphs. In the sense they used it, "graph" denotes a set of points with either a set of lines or a set of arrows. When the points are connected by lines it is called an undirected graph; with arrows it is a directed graph. Erdos and Rényi then ask us to imagine the following experiment. Place ten thousand buttons (our version of points) on the floor. Take a spool of red thread. Break off a piece and choose two buttons at random, then tie them together. Now repeat this process, tying more and more randomly chosen pairs of buttons together. Define a "cluster" as a set of buttons that are directly or indirectly connected. The size of the cluster is just the number of buttons it contains. Imagine picking up a button, and ask how many other buttons you pick up with it.

This is an undirected graph since the threads have no arrow heads. Erdos and Rényi asked how the size of the largest cluster in the graph changes as more and more buttons are randomly interconnected in this graph. The results are deeply surprising. Initially, most buttons are isolated, with a few pairs of connected buttons. As more buttons are connected, little clusters of connected buttons form. Then midsized clusters form. But then something magical occurs. Given enough midsized clusters, adding just a few new connections will connect most or all the midsized clusters into a single giant cluster, called the giant component of the graph. This sudden jump in the size of the largest cluster is, in fact, a "phase transition," much like the phase transition between water and ice. Adding more connections gradually brings the remaining small clusters and isolated buttons to the giant connected component of the graph.

In short, among a set of buttons connected at random by ever more threads, the size of the largest cluster quite suddenly jumps to the point where most buttons are included in it. This phase transition is emergent, is an organizational law of the Laughlin type, and depends upon mathematics, not physics. It plays a key role in the theory of the emergence of collectively autocatalytic sets.

An interesting point is that the details depend upon the total number of buttons. As the number increases, the phase transition becomes sharper, and the size of the giant component becomes a nearly constant fraction of the total number of buttons.

The next step in this explanation is to define a chemical-reaction graph. Consider two monomers—such as the two amino acids alanine and glycine—and call them A and B. Now construct the following picture: Start with the two monomers and imagine reactions that ligate them together, creating AA, AB, BA, and BB. We need to represent the ligation reaction. To do so, we will invent special lines, much like our threads between the buttons. Draw an arrow from A and an arrow from B to a "box," and an arrow from the box to AB. This tripod of arrows into and out of the box, plus the box, represents the given ligation reaction. To indicate that the reaction is not catalyzed, color the tripod arrows black. Draw such tripods and boxes for all the four reactions I just mentioned, from the A and B monomers to all four two-molecule, or dimer, sequences. Once these dimers are created, include in our set of reactions not only ligation reactions, but cleavage reactions breaking the AB dimer into the A and B monomers. Thus the directions of the arrows we drew are only for convenience to show the substrates and product of the reaction in only one direction of the reaction. The actual flow of a chemical reaction—ligation or cleavage—will depend upon its displacement from chemical equilibrium. If there are lots of A and B monomers and no AB dimers, the reaction will tend to flow toward AB in the ligation direction.

We now "grow" the reaction graph. At the next step, continue to add all the possible new molecules that can be formed in a single reaction. Thus, AB and B can ligate to form ABB. AB and BA can join to form ABBA.

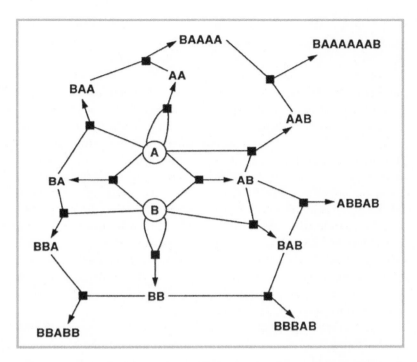

FIGURE 5.1 A hypothetical network of chemical reactions called a reaction graph. Smaller molecules (A and B) are combined to form larger molecules (BA, AB, etc.) that in turn are combined to form still larger molecules (BAB, BBA, BABB, etc.). Simultaneously, these longer molecules are broken down into simple substrates again. For each reaction, a line leads from the two substrates to a square denoting the reaction; an arrow leads from the reaction square to the product. Since reactions are reversible, the use of arrows is meant to distinguish substrates from products only in one direction of the chemical reaction flow. Since the products of some reactions are substrates of further reactions, the result is a web of interlinked reactions.

This picture is a reaction graph. In principle, and perhaps soon in practice, we could ask one or a number of fine organic chemists to take an initial set of N organic molecules and draw the complete reaction graph among all of them in defined conditions of temperature, pressure and so forth.

I want to introduce an idea that will become of the deepest importance in the rest of this book: the adjacent possible. As we will see later,
history itself arises out of the adjacent possible. Consider a reaction
graph with N molecular species, polymer sequences of A, and B
monomers of diverse lengths. Call this initial N the actual. Now ask the
organic chemist to draw all the reactions that these N species might undergo (in conditions defined as above). It may well be that the products
of some of these single-step reactions will not be among the initial N in
the "actual" but will be new molecular species. Call the set of new molecular species reachable in a single-reaction step from the actual, the
adjacent possible. The adjacent possible is perfectly well defined for
chemical-reaction networks. The initial actual plus its adjacent possible
can be considered a new actual, which will then have a new adjacent
possible. Over several iterations, the chemical-reaction graph may persistently reach into ever new adjacent possibles, indefinitely expanding
the total diversity of molecular species in the chemical reaction graph,
unless this interactive expansion is limited, for example, by limiting the
total number of monomers "allowed" per polymer.

The concept of the adjacent possible has physical meaning, which I will
explore more fully when I discuss the nonergodicity of the universe in a
later chapter. Here I note that the early Earth almost certainly had only a
small diversity of organic molecules, perhaps a hundred or a thousand different compounds. Today there are trillions of different organic compounds spread among the roughly 100 million living species. The
biosphere has exploded into its chemical adjacent possible. We will find
similar explosions in economics, human history, and elsewhere. In general,
we have no theories of these explosions, yet they are central to the evolution of complexity. The creativity in the universe is tied to the explosions
into the adjacent possible.

Let's return to the theory of collectively autocatalytic sets as natural
emergent phenomena. Consider a set of organic molecules—say, peptides—as their diversity increases. Write down the reaction graph among
the actual. Ask how the ratio of the number of reactions to the number of
kinds of molecules changes as the diversity of molecules increases. For the
ligation and cleavage reactions above, creating linear polymers, it is easy to

show that this ratio increases linearly, in proportion to the length of the longest polymer. This is easy to see. A linear polymer of length M has $M - 1$ internal peptide bonds. Thus a single M-length peptide can be formed in $M - 1$ ways by ligating other smaller peptides to form it. For example, if the peptide is the sequence AAABBAA, it can be formed by ligating A to AABBAA, or AA to ABBAA, etc. So as length M increases, the ratio of reactions to polymers is roughly M (in fact, it's $M - 2$).

The ratio of reactions to molecules depends on the kinds of reactions we consider. For example, most pairs of organic molecules of modest complexity can react together as two substrates to form two different products in a two-substrate-two-product reaction. Higher-order reactions also occur. Consider a set of modestly complex organic molecules where each pair can undergo at least a single two-substrate-two-product reaction. For N molecules, how many such reactions are there? Well, if there are N distinct molecular species, the number of distinct *pairs* of molecular species is N^2, so there are N^2 reactions among the N molecular species. In each reaction, two substrate species react to form two product species. The ratio of reactions to molecular species is then just N^2 divided by N, or N. Thus, as the diversity of molecular species, N, increases, the ratio of reactions to molecules increases directly proportionally, and the ratio is N. This rate of increase of reactions is much faster than the simple ligation and cleavage reactions noted above, where the ratio increases only with the length of the longest polymer, not with the diversity of molecular species.

Our next step is to ask which of the molecules in the set of N can catalyze (at least at some minimal rate) any of the reactions. In general, we have no current idea what the answer is. Indeed it is exactly this question that we can now ask for libraries of random RNA or peptide molecules. We can also ask the same question for small organic molecules, organometallic molecules, and so forth. It is a clear scientific question.

If we knew which molecules among the N could catalyze which reactions among the N, we could assess whether or not the system contains a collectively autocatalytic set. That, too, is a clear scientific question. But without detailed knowledge of which molecular species catalyze which reactions, researchers have taken three theoretical approaches. In the first,

J. Doyne Farmer, Norman Packard, and I have imagined that each molecule had a fixed probability, P, to catalyze any reaction. This is a too simple model of the distribution of catalysis, but it's a start, and it allows us to ask, at what diversity of molecular species, N, and at what diversity of reactions among them, would we expect a collectively autocatalytic set to emerge.

Let us consider a peptide world with only two monomers, and cleavage and ligation reactions, and the reaction graph among N molecules. We want to see if the system, at each given total diversity of polymers and each probability per polymer of catalyzing each reaction, contains a collectively autocatalytic set. Mathematically, we asked each peptide, does it or does it not catalyze each given reaction? If it does, we draw a blue arrow from that peptide to the box representing that reaction. In addition, we *color* the tripod of arrows into and out of the reaction box *red*, meaning this reaction is *catalyzed*. The red reactions are the catalyzed subgraph of the entire reaction graph.

Having asked our question of each peptide, we can examine the resulting model chemical reaction system to see if it contains a set of molecules that mutually catalyze one another's formation out of some initial exogenously supplied "food" molecules, such as A, B, AA, AB, BA, and BB. Either a collectively autocatalytic set exists in the system or it does not.

Now, the mathematical result that we obtained concerns what happens as the total diversity of molecular species, N, increases. As N increases, the ratio of reactions to molecules also increases. Soon there are very many reactions per molecule. As this happens, more and more reactions among the molecules are catalyzed, and the number of red, catalyzed reactions increases. At some point, so many reactions are catalyzed that a giant catalyzed cluster forms in a phase transition, just as in the buttons and red-thread story. Once the giant cluster forms, the chance that the system contains a collectively autocatalytic set approaches certainty.

The intuitive idea in the theory of the spontaneous emergence of autocatalytic sets is that given a set of molecules, reactions among the molecules, and distribution of catalysis of those reactions by the very same molecules, a diversity of molecular species will be reached, at which time collectively autocatalytic sets spontaneously form in the system. My colleagues and I have

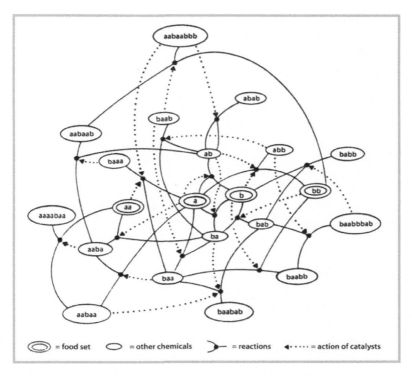

FIGURE 5.2 A typical autocatalytic set built up from a food set, concentric ellipses, consisting in the monomers, a and b, as well as the dimers aa and bb. Catalysis is shown by dotted lines leading from the catalyst to the reaction box, here reduced to a node, representing each of the ligation reactions.

tested this using more-complex models of the distribution of catalytic capacity among the molecules in the reaction system with the same general results. Recently, others have demonstrated mathematically that this phase transition to collectively autocatalytic sets is guaranteed to occur under very general conditions on the distribution of catalysis among the molecules in the system.

That, roughly, is the modern version of the theory of autocatalytic sets. It is perfectly fine mathematically, and shows that for a wide variety of molecular species, reaction types, and catalytic distributions, given a sufficient diversity of molecular species, autocatalytic sets will spontaneously emerge. The diversity of molecular species required to achieve autocatalytic sets depends, more than anything else, upon the complexity of the reactions allowed. Thus a much smaller molecular diversity is needed if two-substrate-two-product reactions are allowed than if only cleavage and ligation are allowed, because in the former case, the ratio of reactions to molecular species increases with species diversity very much more quickly than in the latter case.

The theory of the emergence of collectively autocatalytic sets is plausible. Moreover, we can now test it using libraries of random RNA, peptides, organometallic molecules, or even simpler organic molecules. The amino acid histidine, for instance, catalyzes a variety of reactions, and it is well known that other simple organic molecules catalyze reactions. I hope that in the modestly near future we will be able to test this general model with a variety of classes of organic molecules.

In a very interesting body of work, Walter Fontana, Richard J. Bagley, and J. Doyne Farmer went further and asked whether, in principle, such autocatalytic sets could evolve by heritable variation. The answer is yes. The central idea is this: Imagine a food set of molecules and a collectively autocatalytic set that feeds on them. The molecules in the autocatalytic set are present in higher concentrations than would occur if there were no autocatalytic set. Spontaneous, uncatalyzed reactions among these molecules to create new molecule species will therefore occur at a much higher rate. The first thing to note here is that the autocatalytic set alters the kinetics of the entry of the chemical system into its (for now) uncatalyzed adjacent possible. Fontana et al. realized that one or more of these novel molecules might join the autocatalytic set. The new molecules would then be catalyzed from the autocatalytic set as well, leading to heritable variation. Such systems would be subject to Darwinian natural selection. Thus the emergence of collective autocatalysis, by allowing heritable variation and natural selection, now allows top-down selection on evolving autocatalytic sets as *whole systems*. As different autocatalytic systems evolve and coevolve, the molecular constitution of the universe can be altered.

The very same theory that leads to collective autocatalysis may account for the emergence of a connected catalyzed metabolism. Consider a set of organic molecules on a reaction graph and a set of candidate catalysts—say, organometallic peptides. As the diversity of candidate catalysts increases, ask whether a connected catalyzed set of organic reactions emerges. Then ask whether it leads to the building blocks of the candidate catalysts themselves, such as amino acids or nucleotides. It seems likely that both the formation of collectively autocatalytic sets and a connected metabolism leading to the building blocks can emerge. One always hopes that lipidlike molecules are formed to create bounding membranes that will contain the reactants and that these membranes can grow and divide.

Is the theory of autocatalytic sets reducible to physics? As with Darwin's natural selection, the answer is no, for the autocatalytic theory is agnostic with respect to the kinds of chemicals that might form such sets. More, the theory might be true of the "real world" even if the constants of nature were changed slightly. Thus, the theory cannot obviously be reduced to any specific set of molecules in this specific universe or to some unknown family of similar universes in which autocatalytic sets might arise. The theory of collectively autocatalytic sets is lawlike and transcends any specific physics. This is again the multiple-platform argument. We are again beyond reductionism. Not all laws are reducible to string theory. Some stand at their own level.

Beyond the multiple-platform argument, collectively autocatalytic sets are emergent in a second sense as well: Laughlin's organizational law says that novel properties are mathematical in nature and arise here due to a mathematical phase transition in chemical-reaction graphs as the diversity of the molecular species increases. The theory of collectively autocatalytic sets is a mathematical, organizational "law," independent of physics. Anderson was right. Here Weinberg's explanatory arrows do not point downwards to string theory, but upwards to the collective emergent properties of complex chemical-reaction networks and the organizational, mathematical "law" that may govern the spontaneous emergence of collectively autocatalytic sets.

Finally, there is the interesting "metabolism first" theory. This view holds that metabolism, via uncatalyzed reactions, preceded autocatalysis. Even now there are known chemical-reaction cycles in metabolism that are autocatalytic, in the sense that each passage around the cycle doubles the copy number of at least one molecular species. Beyond this, Albert Eschenmoser, a fine organic chemist, envisions metabolic cycles whose members act as actual catalysts to create a collectively autocatalytic metabolism. Robert Shapiro, in a June 2007 article in *Scientific American*, puts forward a similar thesis, with the addition that a source of free energy in an exergonic reaction is coupled to and drives the metabolic autocatalytic cycle. This new area of "systems chemistry" is likely to see fine experimental work and may in fact achieve collectively autocatalytic sets of organic molecules. It is not at all impossible that a "metabolism first" theory could lead to the synthesis of amino acids, which could then form the organometallic or peptide catalysts that create a giant component in a catalyzed-reaction graph, including the metabolites and their reactions; the formation of lipids to make a bounding, budding membrane; and the peptides in the collectively autocatalytic system that now has a metabolism and is a collectively autocatalytic whole.

What are we to make of the efforts to date in attempting to understand the origin of life on Earth and elsewhere in the cosmos? First, attempts to obtain molecular reproduction by template replication of RNA have so far failed. The RNA world in the restricted sense of a *single* self-reproducing molecule has yet to achieve molecular reproduction, but it may evolve a ribozyme polymerase able to copy itself. It is not clear that such a molecule would be evolutionarily stable in the face of spontaneous mutations of the ribozyme and the accumulation of ever more erroneous copies. Nor does the single RNA polymerase version of RNA world theory explain why modern cells are complex, collectively autocatalytic wholes. The more general view of the RNA world, enunciated by Walter Gilbert, envisions a system of RNA ribozymes that form one another and is wholly consistent with a theory of the emergence of collectively autocatalytic sets as discussed above. The only examples of molecular reproduction we have are single-stranded DNA and peptides singly and in collectively autocatalytic sets. Both have been conclusively demonstrated experimentally. Because peptides can form single autocatalytic reactions and also collectively autocatalytic sets, molecular reproduction definitely does not depend on the symmetry of DNA and RNA double helix molecules or on template replication. Collectively autocatalytic peptide sets are chemically real.

Still, it remains to be proved that the general theory of collective autocatalytic sets applies to real molecules. The theory *does* explain why molecular self-reproduction is complex and whole, for collectively autocatalytic sets contain, from the outset, a minimal diversity of mutually catalytic molecules. Such sets require that the components be confined to a small volume so they can interact with one another. A more adequate theory, and supporting experimental work, would demonstrate that collectively autocatalytic sets arise spontaneously in complex chemical-reaction networks, and that the molecules would catalyze a minimal metabolism to support the synthesis of the polymers in the system, plus lipids to form a bounding membrane.

But we can say at a minimum that it is scientifically plausible that life arose from nonlife, probably here on Earth. It is also plausible that we will succeed in creating modestly complex self-reproducing chemical nonequilibrium reaction systems capable of heritable variation and natural selection. This would be a major scientific triumph. While the definition of life is subject to debate, it seems likely that we will create life by some reasonable definition sometime in the coming century.

Further, we will create a general biology linking physics, chemistry, mathematics, and biology anywhere in the universe. Since alien life-forms, evolving by natural selection, will have functions (like the pumping of blood by the heart) that are subsets of their causal features, this general biology will not be reducible to physics alone. It lies beyond reductionism. Life has emerged in the universe without requiring special intervention from a Creator God. Should that fact lessen our wonder at the emergence and evolution of life and the evolution of the biosphere? No. Since we hold life to be sacred, we are stepping towards the reinvention of the sacred as the creativity in nature.

6

AGENCY, VALUE, AND MEANING

If reductionism leaves us as nothing but particles in motion, then only brute facts exist in reality, or, as philosophers sometimes say, in the "furniture of the universe." But the agency that arises with life brings value, meaning, and action into the universe. The existence of agency takes us beyond reductionism to a broader scientific worldview.

I will start with human agency—the root of human action. But my larger purpose is to trace agency to its earliest roots at or near the origin of life itself, in what I will call a minimal molecular autonomous agent. The agency of early life is a kind of protoagency, far short of full human agency, but presumably the rudiment out of which human agency evolved. In it I believe we can find the origin of action, meaning, doing, and value as emergent realities in the universe.

AGENCY AND TELEOLOGICAL LANGUAGE

If you are reading this book, you are a full-fledged human agent. For example, you are currently reading this sentence. You probably had breakfast, lunch, and dinner in the past day or two. You may have driven your car to the market to buy groceries. In all these commonplace activities, you are, in a general sense, acting on your own behalf. Your actions are literally altering the physical universe. Think not? When the United States sent a rocket to Mars, human action altered Mars's mass, subtly changing the celestial dynamics of (at least) the solar system. We only have to fly over much of the planet to see how we have altered the native face of this world.

We are agents who alter the unfolding of the universe. Philip Anderson wrote charmingly about agency, "If you doubt agency, tell your dog to 'come' and watch the agonized indecision on his face." Our own dog, Winsor (recently lost to cancer), used to illustrate Anderson's point on an hourly basis. Winsor's response when I told him to come as he was busy smelling a bush was perfect: a sidelong glance at me, as if thinking, "I've got time here." Then back to sniffing the bush. "Come!" Another sidelong glance, and more sniffing. "Winsor, *come!!*" A reluctant trot to my side. Some would say I am anthropomorphizing, but I do not think so. There are too many stories of apparently conscious animal behavior, including a sense of fairness in higher primates, for this skeptical stance to be convincing.

Another example made the local newspaper in Edmonton, Alberta. A grandmother left her infant grandson in a small rocker in the backyard with their dog. A rattlesnake coiled to strike the baby. The dog threw itself between the baby and the snake, was bitten seven times, and, happily, survived. Agency, I believe. I truly believe, as does virtually anyone who has lived with a dog, that Winsor knew perfectly well what I was doing, and was gauging his response based on his intimate familiarity with my moods. I am completely persuaded that Winsor was an agent, as is a tiger chasing a gazelle, and the gazelle fleeing in terror. So was T. rex chasing prey such as a hadrosaur. Winsor liked to ride in our car between the front seats, looking out the front window. Many times, as we approached the sharp bend in the road approaching our house, he began to lean into the curve—well before the car started its turn. Similarly, the tiger alters its chase as the gazelle alters its race for life.

I mention agency in animals because I want to stress that agency reaches beyond humans. We do not know its biological limits. I will, however, try in this chapter to find the minimal physical system, which I will call a molecular autonomous agent, that we might be willing to call an agent.

In physics there are only happenings, no doings. In biology there are functions, like that of the heart pumping blood, that are emergent with respect to physics. As we saw, the physicist cannot see function: when looking at an animal heart he cannot pick out the functional subset of processes, in this case pumping blood, that arose due to natural selection from the myriad other causal properties of the heart. With agency, values have emerged in the universe. With value comes meaning, not yet in the semantic sense, but meaning in the sense of mattering to some entity.

Another point is immediately worth noting. An action, in similarity to the function of the heart, is only a subset of the causal events that occur amid the full set of happenings surrounding the action. For example, if you lift a glass of water to drink a sip, the precise pathway that your arm takes is largely incidental to the act. If you drive to the grocery store to shop, the precise street route you take is incidental. This may again bear on reducibility, for there seems no way, were a physicist able to deduce all the events involved in your going to the grocery store, for him to pick out the *relevant* subset of happenings that constitute the action. Again, the relevant features of actions, like biological functions arising due to evolution, cannot be discriminated by the physicist, for whom there is only the totality of what happens in the universe.

To discuss molecular autonomous agents we first need to consider teleological language. When we talk about fully human action, we normally attribute it to conscious reasons, motives, intentions, purposes, and understanding. Such language is teleological. These issues are of practical importance in our everyday interactions with one another, as well as in courts of law. Aristotle himself raised the distinction between acting and acting under compulsion. He considers a ship captain in a storm who decides to throw much of the cargo overboard to save the ship. Was the captain sufficiently compelled by the storm that throwing the cargo overboard was not his action? Aristotle narrowly comes down on the side of attributing the action to the captain. But consider the case of my intending to turn on the light in a room. I carry out what modern philosophers of mind sometimes call a basic act: I move my finger against the switch so that the light is turned on. But suppose, to my surprise, the sudden light startles a man in the room, who has a heart attack and dies. Did I kill him? Most of us would say that the man's death was a consequence of my act, but one that would be unanticipated by a reasonable person. My act, therefore (by the reasonable-person test), is limited to turning on the light. As a number of philosophers of mind in the past forty years have noted, intentions are very closely associated with action. The act for which I am morally responsible is turning on the light, not killing the man. Philosopher John Searle attempted to convey the "closeness" of intention and understanding to action by saying that actions are events occurring "under a description" (i.e., the intention). Whether Searle captures this closeness appropriately or not, it is worth bearing this closeness in mind.

Let me offer the familiar philosophical and scientific caveats: the use of teleological language to describe reasons, intentions, and understandings is not scientific. Only you have personal access to knowledge of your intentions. And most commonly: *teleological language, framed in terms of reasons, purposes, and intentions, can be "cashed out," or turned into a fully causal account of what has occurred, say, in your brain's neurons.* Therefore, the "scientific," objective account of your actions is found in the eventual neural-causal account.

The last point is debatable. The certainty that we can cash out a teleological description in a causal account takes roughly the following form: Billiard balls bounce off one another, and we can give a causal account of their dynamics. A human being is nothing but a complex machine, and ultimately we can give a similar causal account of, for example, nerve impulses in brain regions giving rise to specific behaviors. In this argument, one needs to be clear what a "causal account" is supposed to be. As with the moving billiard balls, it is supposed to point to a web of events that are jointly necessary and sufficient to determine the event we wish to "explain": my act as an agent. To be successful, that causal account must also account for the perceived closeness between intentions and actions—what Searle called acting "under a description." At present there seems no way to account for this closeness. Suppose I go to the hardware store for a new light switch (one with a dimmer that I can turn on gradually). We might possibly obtain a full causal account—in terms of, say, neural activities—of all the events that happen in and through my body, including driving to the store and my elbow hanging out the car window. But there does not seem to be any way to pick out from this full account the relevant aspects of these events, that is, the subset of events that constitute my action of going to the store. This problem becomes even more severe if we imagine deducing all the events that happen in and through my body from fundamental particle physics (a deduction that, in any case, cannot be carried out).

These difficulties are yet another version of the multiple-platform argument. Just as we cannot replace a statement about computing with a finite set of prespecified statements about any specific physical system, or a statement about natural selection with a finite set of pre-specified statements about events in this or other universes, so we cannot replace the statement "Kauffman went to the hardware store" with a finite, prespecified set of statements about the physical events that constitute my going to the store. Instead of driving there I might have walked, run, skipped, parachuted, or gone by a scooter, along myriad pathways. We cannot reduce action statements to physical ones. Aristotle himself noted this feature of organisms acting. He posited four kinds of causes, one of which was "telos," which means purpose or "end," to cover actions.

In short, teleological language cannot be replaced by physical language for two reasons. We have no way to pick out the relevant subset of events that constitute the action, nor can we prespecify a finite, necessary, and sufficient set of physical events that jointly characterize the action statement. A philosophical position called reductive elimination holds that such a replacement is needed so that the action statement is just a "shorthand" description for the more accurate physical statement. If there can be no such replacement, then teleological language, the language of agency, cannot be eliminated and replaced by a physical account of events. Teleological language is beyond reductionism.

The difficulty eliminating such language is that teleological language is what philosopher Ludwig Wittgenstein called a "language game." It has turned out, largely by Wittgenstein's own seminal work on language games, that eliminative reductionism, just noted above, in general fails. I will give two examples.

Consider the statement, "England has gone to war with Germany." We all understand this. Now let us try to give a set of necessary and sufficient conditions about human actions for England and Germany to come to be at war. Well, the prime minister might say in Parliament, "I move we declare war on Germany." A vote is taken, passed, England declares war on Germany, and Germany responds with a similar declaration. Alternatively Queen Elizabeth might enter Parliament and cry, "Ladies and chaps, after the Germans!" And so England and Germany go to war. Or ten thousand angry soccer fans in Leeds might pile into rowboats with pitchforks, cross the Channel, paddle up the Rhine, and attack astonished German soccer fans. Thus there are no prespecifiable lower-level statements about human actions that are necessary and sufficient to describe England and Germany going to war. "England and Germany go to war" cannot be replaced by a prespecified list of statements about simple human actions.

Similarly, consider a statement like "Smith was found guilty of murder in Nebraska's superior court on May 5, 2006." Think of the interwoven concepts required for you to understand this statement: guilt and innocence, evidence, admissible evidence, jury, swearing in, perjury, legal responsibility, and more. This mutually defining circle of concepts, jointly needed for one another, are what Wittgenstein meant by a language game. We cannot, he claimed, translate out of this language game to any necessary and sufficient set of statements about simple human actions devoid of this legal circle of interdefined concepts. That is, we cannot reduce statements about trials at law to statements about normal human actions. Wittgenstein would say here that the reduction simply fails.

If Wittgenstein is right, as almost all philosophers believe, then we cannot deduce algorithmically the language game of law and guilt from the language game about intentional human actions. But that means that the mind, in grasping the legal language game, is not carrying out an algorithm, where, again, an algorithm is an "effective procedure" to produce an output given an input. There appears to be no effective procedure to get from nonlegal descriptions of human actions to legal descriptions. When we get to the chapter on mind, this shall be one of the pieces of evidence that the human mind need not act algorithmically, for we understand legal language but cannot deduce it from descriptions of other human actions, again if those other human actions are devoid of the interdefined circle of legal concepts.

With respect to teleological language, the same language-game argument works. As we have seen, eliminative reductionism, replacing teleological statements by a finitely prespecified set of statements about physical causal events, say in neurons, cannot be carried out. Thus, eliminative reductionism fails.

In chapter 13, The Quantum Brain?, I will return to human agency, free will, and the moral responsibility that presumes teleological language.

MINIMAL MOLECULAR AUTONOMOUS AGENTS

We seek the origin of agency, and with it the origins of meaning, value, and doing. To set ourselves this task is to accept that agency, meaning, value, and doing are real parts of the universe and are not reducible to physics, in which only happenings occur. What is the simplest system to which we might be willing to apply teleological language? When biologists talk of a bacterium swimming up a glucose gradient "to get" sugar, they are using teleological language. But we would not say a ball was rolling "to get" downhill.

Teleological language becomes appropriate at some point in the tree of life. Let us stretch and say it is appropriate to apply it to the bacterium. We may do so without attributing consciousness to the bacterium. My purpose in attributing actions (or perhaps better, protoactions) to a bacterium is to try to trace the origin of action, value, and meaning as close as I can to the origin of life itself.

But a bacterium is a very complex cell. Is there a simpler system to which we would be willing to apply teleological language? Let me call the simplest such system a minimal molecular autonomous agent. By "autonomous" I mean that the agent can act on its own behalf in an environment, like the bacterium swimming up the glucose gradient or me going to the hardware store.

In *Investigations,* I offered a tentative definition of a minimal molecular autonomous agent. Such a system should be self-reproducing and should carry out at least one thermodynamic work cycle. (I will discuss work cycles in a moment.) With the help of philosopher Philip Clayton, I have amended this definition to add the requirements that the agent be enclosed in a bounding membrane, have one or more "receptors" able to detect food or poison, and respond, typically via work like swimming up the glucose gradient, to get the food or avoid the poison. Thus the agent must be able to detect, make a choice, and act. Virtually all contemporary cells fulfill this expanded definition. Such systems are emergent in Laughlin's sense, with novel properties, for they are agents and can act, where the action is the relevant subset of the causal events that occur. They can evolve by heritable variation and natural selection that assembles the molecular systems in the cell that allow response to food or poison, typically by doing work.

To understand agency, then, we need to understand both self-reproduction, the topic of the last chapter, and work cycles. The most convenient way to gain insight into this subject is through the famous Carnot thermodynamic work cycle. The system consists of two temperature reservoirs, hot and cold. In addition, there is a cylinder with a piston in it, and a compressible working gas between the cylinder head and the piston. The system starts with the gas compressed and hot. The cylinder is placed in contact with the hot reservoir and the working gas is allowed to expand to move the piston down the cylinder in the first part of the power stroke. Expansion will make the gas cool, but contact with the hot reservoir will keep it near its initial hot state. Halfway down the power stroke, the cylinder is moved out of contact with the hot cylinder. The working gas continues to expand, pushing the piston further down the cylinder in the remainder of the power stroke, until now cooling gas reaches the temperature of the cold reservoir. This power stroke is what physicists call a spontaneous process, one that occurs without the addition of energy. A major discovery of nineteenth-century physics was the distinction between spontaneous and nonspontaneous processes; the latter can't happen unless work is done on them. An example of a nonspontaneous process is rolling a ball up a hill. A spontaneous process would be the ball rolling down the hill. The complete work cycle combines the spontaneous power stroke with a nonspontaneous process that pushes the piston back up the cylinder and recompresses the working gas. This recompression requires outside work to push on the cylinder.

In short, the Carnot cycle, like all engines, works by alternating spontaneous and nonspontaneous processes. At the end of the power stroke, the engine cannot repeat the power stroke and cause motion. Instead, work must be done to push the piston up the cylinder and reheat the now cool gas. So outside force is used to push the piston back up the cylinder. The secret of the Carnot engine, and the steam engine, is that it takes less work to recompress a cold gas than a hot gas. So at the beginning of the recompression stage, the cylinder is moved into contact with the cold reservoir. As the piston is pushed up the cylinder, the working gas tends to heat up, but this heat diffuses into the cold reservoir, keeping the gas cool and easy to recompress. Partway up the recompression stroke, the cylinder is moved out of contact with the cold reservoir, work is done to continue to push the piston up the cylinder, and now the working gas reheats, reaching its initial temperature just as the piston returns to its starting position.

The Carnot cycle tells us a number of deeply important facts about work cycles and, by extension, about agency. First, work cycles cannot occur at equilibrium. Since agency requires the performance of at least one work cycle, it is inherently a nonequilibrium concept. Second, work cycles link spontaneous and nonspontaneous processes. The chemical equivalent is simply stated. A chemical reaction will go "downhill" in "free energy" to equilibrium (where the rate of conversion from X to Y equals the rate of conversion from Y to X). Pushing the chemical system away from equilibrium—say, to an excess of X—is a nonspontaneous process that requires work. This is the chemical equivalent of rolling the ball uphill. It is a requirement of autonomous agents that they maintain an internal environment that is displaced from chemical equilibrium. We do so by eating and excreting.

Third, the work cycle is, in fact, a cycle of processes. So is the cell cycle. Were the engine not a cyclic process, energy would flow downhill and then simply stop, unable to organize the processes needed to get itself back up the hill. In *Investigations*, I describe a whimsical Rube Goldberg machine in which a ball fired from a cannon strikes a paddle wheel straddling a well. The paddle wheel is caused to spin, winding up a rope leading

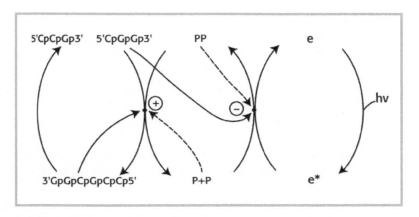

FIGURE 6.1   A first hypothetical molecular autonomous agent. The entire system is an open non-equilibrium thermodynamic system driven by two sustained sources of food. The sources of food are two singled-stranded DNA trimers and a photon source, hv. The two single-stranded DNA trimers undergo a ligation reaction to form a second copy of the hexamer, thereby replicating the hexamer. There are two reaction pathways of synthesis of the hexamer and its breakdown back to trimers, one direct, the other that is coupled to pyrophosphate, PP, breakdown. The forward pathway is coupled to the exergonic breakdown of pyrophosphate, PP, as are two monophosphates, P + P. The *exergonic–spontaneous* drop in free energy from the breakdown of PP drives *excess, hence endergonic,* synthesis, hence *excess replication* of the hexamer, compared to the equilibrium otherwise achievable. PP, in turn, is reformed from relegation of P + P by an *endergonic* reaction that is driven by coupling to a photon-excited electron, e*, which gives up free energy in an exergonic reaction to return to a ground state, e, while adding that energy to the endergonic synthesis of PP from P + P. The electron at the ground state, e, is reexcited endergonically by absorption of another photon, hv. The hexamer catalyzes the reaction leading to its own excess synthesis. One of the trimers catalyzes the reaction leading to PP resynthesis. The activation of the hexamer enzyme activity by monophosphate and inhibition of trimer enzyme activity by pyrophosphate ensures that the synthesis of hexamer and restoration of pyrophosphate occur *reciprocally.*

to a bucket in the well filled with water. The bucket is pulled up, twists over the paddle wheel, and pours its water into a funnel tube that flows downhill to open a flap valve that lets water flow onto my growing bean field. I love my machine. It links lots of spontaneous and nonspontaneous processes, and is magnificent for what I call "propagating work." But once it has run its course, it cannot restart. Someone has to reset the cannon, powder, cannonball, paddle wheel, rope, bucket, and so forth. Notice that this resetting requires work: nonspontaneous processes resetting all the parts of the machine, cannonball, pail of water in the well, and so forth. My machine cannot continue to function by itself without further work done on it to complete a cycle and reset it to its initial state. The Carnot engine can continue to function because it works cyclically.

Fourth, after my noncyclic Rube Goldberg device has watered my bean field and has not been reset, notice that the mere addition of new energy, here powder in the cannon, cannot lead to more work being done. The cannonball has not been replaced in the cannon, the bucket has not been unwound from the paddle wheel and lowered into the well and filled with water. *Without a cyclic process, the Rube Goldberg system linking spontaneous and nonspontaneous processes cannot make use of an outside energy source.* This is a critical issue. Erwin Schrödinger and all subsequent biologists have recognized that cells require negentropy, i.e., food and energy. *What has not been noticed is that, if spontaneous and nonspontaneous processes are linked together as in my Rube Goldberg, then without a self-resetting work cycle linking spontaneous and nonspontaneous processes, that food energy cannot be used to perform more work.*

Let's go back to my Carnot cycle to address the mechanisms that move the cylinder into and out of contact with its two heat reservoirs. They, too, are part of the cyclic operation, part of the propagating organization of process. In a real engine, cams, gears, and other devices move the cylinder into and out of contact with the hot and cold reservoirs. In the hypothetical molecular autonomous agent in figure 6.1, the molecules that activate and inactivate the enzymes of the system play just this role in coordinating the agent's work cycle. Similar couplings coordinate work cycle processes in cells.

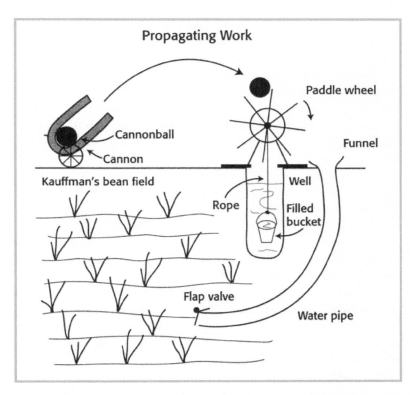

FIGURE 6.2 *Propagating work* takes place via my outstanding Rube Goldberg invention. The cannon fires a cannonball that hits the paddles of the paddle wheel that I built, which is beside a well. I have attached a red rope to the axle of the paddle wheel and tied the other end to a bucket, which I lowered in the well so it filled with water. The spinning paddle wheel winds the rope up and raises the bucket, which tips over the axle, pouring water into the funnel leading to the water pipe. The flowing water opens the flap valve I built into the end of the pipe, thereby watering my bean field. Note that my *machine cannot "work again"* unless I *reset the parts of it*—for example, by placing more powder in the cannon, lowering the bucket into the well again, and replacing the cannonball in the cannon, then firing the cannonball again, I can water my bean field again.

In addition to performing work, the work cycle can store energy. Thus if excess X is synthesized by doing work, that excess energy can later be used for error correction in a cell (a point made to me by Philip Anderson and Daniel Cloud, who also pointed to the selective advantage of storing energy in a variable environment where food supplies may be irregular.) For example, you store energy in fat and in linked sugars that form polysaccharides. Building these molecules requires thermodynamic work. In the absence of food, you break them down to obtain the small molecules, such as glucose, that you require for metabolism.

I began this discussion by postulating that autonomous agents are self-reproducing and perform at least one work cycle. Is it of selective advantage to bother linking spontaneous and nonspontaneous processes into a web of propagating work—either my acyclic Rube Goldberg version or a cyclic version that resets itself to enable it to accept more energy? The answer may be simple, surprising, and fundamental to the diversity of the biosphere.

Let me introduce here a couple of specialized terms. A spontaneous process that does not require an input of work—such as a ball rolling downhill—is called *exergonic*. A nonspontaneous process—one that does require an input of work—is called *endergonic*. Ghadiri and von Kiedrowski have made entirely exergonic self-reproducing molecular systems. It's reasonable to suppose that the earliest self-reproducing molecular systems were like this: able to replicate themselves in some process chemically equivalent to a ball rolling down a hill, but not able to reset themselves to do it again. This ability, I would argue, falls short of agency, for it does not involve a work cycle.

Suppose life were limited to such systems and could not link exergonic to endergonic processes. For instance, it takes work by the cell to create ATP, the main form in which cells store energy. What if this chemical pathway were unavailable? Most endergonic synthetic processes in the cell would cease. *In short, if only linked exergonic processes could occur, the total diversity of linked processes and events that could arise would be very much more limited than it is if exergonic and endergonic processes can be linked together.* In real cells, spontaneous and nonspontaneous exergonic and endergonic processes are vastly webbed together so that sunlight yields the construction of redwood trees and other plants eaten by herbivores, which are eaten by carnivores, which are eaten by scavengers and bacteria. It is crucial for a cell in a variable environment and coevolving with other cells to increase the diversity of what it can do, hence linking exergonic and endergonic processes is an enormous selective advantage. *But this linking of exergonic and endergonic processes into a web of processes,* like my cannon-and-paddle-wheel beanfield irrigation device, *requires that the system have the capacity to be reset so it can accept an energy input and work again. Work cycles linking exergonic and endergonic processes arose for good selective reasons. Thus, if agency requires molecular self-reproduction and performance of one or more work cycles, the emergence by selection of agency, given self-reproduction and heritable variation, was of selective advantage. Agency is, then, natural and expected.*

In short, we can define agency as the uniting of molecular reproduction with work cycles. It arose, we may assume, because it offered a vast selective advantage in a variable environment over a purely exergonic self-reproducing chemical system like Ghadiri's or von Kiedrowski's. If a protocell is to evolve to perform a variety of tasks in a complex variable environment, linking spontaneous and nonspontaneous processes increases the diversity of "what can be done." But this linkage requires resetting, and so requires closure into work cycles. Agency almost had to arise in evolution.

It may be that molecular autonomous agents provide a minimal definition of life. If so, viruses are not alive, for they do not do work cycles. Alternatively, it may be that agency distinguishes organisms from mere viral life. In their efforts to create life, scientists are also unwittingly working to create protoagency.

AGENCY, MEANING, VALUE, DOING, PURPOSE

With agency, at whatever level of evolution we are willing to recognize it, meaning, values, doing, and purposes emerge in the universe. Earlier in this chapter, I discussed the use of teleological language in the case of human agency. I believe the same language is legitimate with respect to bacteria and minimal molecular autonomous agents. We have already seen that biological function, such as the heart pumping blood, is distinct from other (nonfunctional) causal consequences, such as heart sounds, by the fact that the organization of processes and structures we call a heart came into existence by virtue of natural selection for its capacity to pump blood. Evolution by natural selection, acting on heritable variants, is what creates the distinction between functional causal features and side-effect causal features of organisms.

I also discussed the irreducibility of Darwinian natural selection to physics. We cannot write down the laws of the evolution of the biosphere. Nor can we simulate the evolution of our specific biosphere, because we can neither carry out the infinity of simulations that would be required nor confirm which is correct with respect to the throws of the quantum dice that occurred in our actual biosphere. In addition, the principle of natural selection can apply to many versions of life capable of heritable variation. So natural selection cannot be reduced to any specific physical basis—the philosophical multiple-platform argument.

Thus the heart is both epistemologically and ontologically emergent. It is epistemologically emergent because we cannot deduce it from physics alone. It is ontologically emergent because the very existence of its specific organization of structures and processes in the universe was assembled by heritable variation and natural selection, which cannot be reduced to physics.

Consider then a bacterium swimming up the glucose gradient. The biological function that is being fulfilled is obtaining food. The capacity to do so by detecting the local glucose gradient and swimming up it was assembled into a working organization of structures and processes by natural selection. This requires at least one receptor for glucose to discriminate between the presence and absence of glucose, or better, two receptors spaced some distance apart to detect the presence or absence of a steep enough local glucose gradient. Without attributing consciousness to the bacterium, we can see in this capacity the evolutionary onset of choice and thus of meaning, value, doing and purpose. The technical word for meaning is *semiosis,* where a sign means something. Here, the bacterium detects a local glucose gradient, which is a sign of more glucose in some direction. By altering its behavior and swimming up the gradient, the bacterium is *interpreting* the sign. The bacterium may, of course, be *mistaken*. Perhaps there is not much glucose to be found in that direction. Neither "signs," "interpretation," nor "mistakes" are logically possible in physics, where only happenings occur. Thus *meaning* has entered the universe: the local glucose gradient is a *sign* that means glucose is—probably—nearby. Because natural selection has assembled the propagating organization of structures and processes that lead to swimming up the glucose gradient for good selective reasons, glucose has *value* to the bacterium. And because getting food is the function of this organized behavior, as assembled by natural selection acting on fitter variants, getting food is the *purpose* of the activity, and is the *doing* or *action* of the bacterium. Just as heart sounds are not the function of the heart, other causal features of the swimming bacterium and its precise route up the gradient are side effects with respect to the purpose of the activity.

There is a final point that will become a central issue in chapter 17, on ethics. The Scottish Enlightenment's skeptic philosopher David Hume wrote that one cannot deduce "ought" from "is." That is, he said, from what "is" the case, there is no logical connection to what "ought" to be the case. Hume's critique set the stage for modern ethics, from Kant to the utilitarians to today. Was he right? Yes and no. In physics, there are only happenings, only facts, only statements of fact about what "is." But we have seen that biology is not reducible to physics, nor is agency, an aspect of life, reducible to physics. With agency, with the bacterium swimming up the glucose gradient, values enter. Once this is true, meaning and "ought" enter the universe. To say it simply, in order to get the glucose, the bacterium "ought" to swim up the glucose gradient. The details of its swimming, note, do not matter. But this agency-borne "ought" is, we have seen, not reducible to physics, and the language of bare facts, of what "is." Thus, "ought" is also emergent, and not reducible to statements restricted to "is." So Hume is both right—we cannot deduce "ought" from "is"—and not right. Values, meaning, doing, action, and "ought" are real parts of the furniture of the universe. "Ought" is central to much of human action and all our moral reasoning.

All this, too, is part of reinventing the sacred.

7

The Cycle of Work

An organized being is then not a mere machine, for that has merely moving power, but it possesses in itself formative power of a self-propagating kind, which it communicates to its materials though they have it not of themselves; it organizes them, in fact, and this cannot be explained by the mere mechanical faculty of motion.

—*Immanuel Kant*

This quotation, from Kant's *Critique of Judgment*, points to the subject matter of this chapter: propagating organization of processes. As we will see, the phenomena are real and right before our eyes. Nevertheless, we appear to lack an adequate theory of what we are looking at. It is said that the Native Americans witnessing the arrival of Columbus did not "see" his ships, for they had no categories of thought in which to place ships. Such is our situation now. Sometimes science advances by seeing things anew. The evolution of life has yielded astonishing organization that links complex, spontaneous and nonspontaneous processes of enormous intricacy. Part of our natural astonishment at life stems from this outrageous, complex organization of process. To the devout who require that a Creator God have brought it forth, science says, wait—we are coming to understand how it all arose naturally with no Creator's hand.

With the wide flowering of mathematics, physics, chemistry, biochemistry, biophysics, and biology, you would think that we would already have the scientific concepts in hand to understand the life that lies before us, but we do not. The manifold web of organized processes we call the biosphere largely reflects natural selection and the evolution of this specific biosphere and hence, while not contravening physical laws, is not reducible to physics. In this chapter I explore a web of concepts, hinted at by Kant, that concern propagating organization of process. From the larger perspective of this book that seeks to move beyond reductionism to a scientific worldview of emergence and ceaseless creativity, the evolutionary emergence of overwhelming organization of process is the most visible example of a phenomenon that is in no way accessible to reductionism.

For the nonscientific reader, the present chapter is an example of the struggle in science to formulate questions that are still unclear. The attempt to "find" the needed concepts for propagating organization is an example of the requirement for imagination and even wonder in science. There appears to be no algorithm, or effective procedure, to find the concepts that we need. We do in fact live our lives forward, often without knowing, which requires all of our evolved humanity, not just "knowledge." We truly must reunite that which the metaphysical poets split asunder. The very attempt to articulate a scientific question is an example of this aspect of our humanity.

The rise of molecular biology in the twentieth century brought to the study of living systems a focus on "information," a concept that manages to be both restrictive and unclear. We have been enjoined to more or less ignore energy, work, and other properties of cells and focus on the "information master molecules of life," the genes, and the processes that genes control: transcription of RNA, translation to proteins, the differentiation of cells as they develop from the fertilized egg, and so forth. But "information," particularly the one-way flow of information from DNA to RNA to proteins that remains molecular biology's central dogma, is an inadequate framework for describing much of what we see in the cell. Cell biologists, for instance, have been struggling to understand the behavior of organelles called mitotic spindles, where molecular motors attach to one or more molecules called microtubules, themselves able to grow and shrink. The motors race back and forth holding onto the microtubules, and upon reaching the ends, gather the microtubles into a variety of complex mechanical structures that, among other things, form the mitotic spindles that move daughter chromosomes in sets to the two daughter dividing cells such that each gets a full complement of chromosomes. Here it is clear that mechanical *work* is done, for the chromosomes are moved bodily to the appropriate daughter cells. More broadly, cells do some combination of mechanical, chemical, electrochemical, and other work and work cycles in a web of propagating organization of processes that often link spontaneous and nonspontaneous processes. But the sorts of questions asked by molecular biology, with its emphasis on how one molecule "instructs" another, are of little help in understanding this web of propagating organization of processes and these work cycles.

If we ask what "work" is, it becomes surprisingly difficult to define. To a physicist, work is force acting through a distance (such as pushing a hockey stick and accelerating a hockey puck). But chemist Peter Atkins says work is more than this. It is the constrained release of energy into a few degrees of freedom. We have already seen this in the Carnot engine with the working gas in the cylinder pushing down on the piston in its power stroke. Each of the randomly moving gas molecules in the working gas enjoys many degrees of freedom—many possible combinations of speed and direction—but when they are confined between the cylinder head and the piston they exert a pressure on the piston, constraining that random motion into the few degrees of freedom represented by the motion of the piston down the cylinder. Work has been done on the piston by the constrained expanding working gas.

Now let's ask what the constraints on the release of the working gas energy are. Obviously, the constraints are the cylinder and the piston located inside it, with the working gas located between the head of the cylinder and the face of the piston.

In analyzing such a system, physicists since Newton have put in the constraints (defined in this case by the surfaces of the cylinder and piston that contain the gas) "by hand" as what are called the mathematical "boundary conditions" on a system, rather like the boundaries of a billiard table that keep the balls from rolling off to infinity or under the sofa. Given boundary conditions, physicists state the initial conditions, particles, and forces, and solve the equations for the subsequent dynamics—here, the motion of the piston.

But in the real universe we can ask, "Where do the constraints themselves come from?" In the case of the cylinder and piston, the answer is simple and startling. It takes real work to construct the cylinder and the piston, place one inside the other, and then inject the gas.

Thus the release of energy must be constrained into a few degrees of freedom to create work. And it takes work itself to construct those very constraints on the release of energy that then constitutes further work! This violates no known law of physics. But I think it is a major clue to something we cannot yet see, something big that is right in front of us: I will call it the propagating organization of process. It takes work to constrain the release of energy, which, when released, constitutes work. No constraints, no work. No work, no constraints. This is not taught in physics, but it is part of what cells do when they propagate organization of process. They have evolved to do work to construct constraints on the release of energy that in turn does further work, including the construction of many things such as microtubules, but also construction of more constraints on the release of energy, which, when released, does further work to construct many things, including more constraints on the release of energy. Thus a self-propagating organization of processes arises in cells.

In the previous chapter I sketched my playful Rube Goldberg device with the cannon, powder, cannonball, paddle wheel straddling the well, and rope down to the water-filled bucket. The fired cannonball hit the paddle wheel, spun it, wound up the rope, and spilled the water into a funnel tube to flow down and water my beanfield. In this case, it is I who have constructed all the mechanisms with their boundary conditions, positioning the cannon, building the paddle wheel, tying the rope with the filled water bucket, building the funnel tube, and leading it to my bean field. I myself did the work to arrange the constraints on the release of energy. The actual events that result mix spontaneous and nonspontaneous processes. But nowhere do these processes themselves construct constraints on the release of energy. Yet cells carry out work to construct constraints on the release of energy and the resulting work does in fact construct many things, including further constraints on the release of energy that, doing yet more work, can, in turn, construct yet further things including more constraints on the release of energy. If constraints can be thought of as boundary conditions, then cells build their own boundary conditions. Indeed, they build a richly interwoven web of boundary conditions that further constrains the release of energy so as to build yet more boundary conditions.

We have absolutely no clear picture of this, no theory of it, not even an idea of what such a theory might "look like." I ask nonscientists and scientists alike to feel the humbling yet ennobling mystery (or, if you prefer, the confusion) when we cannot see what is right in front of us.

Perhaps the simplest physical system that does something of what I am talking about is a river. The riverbed constrains the flowing water, acting as the boundary conditions on its flow. But meanwhile, the flow of the river erodes and changes the riverbed. So this familiar system releases energy into a few degrees of freedom—water flowing down the gravitational potential—and simultaneously does work, erosion, to modify its own boundary conditions. Fine, an example helps, but we have as yet no theory for systems that do work and work cycles to build their own boundary conditions that thereafter modify the work that is done that then modifies the boundary conditions in the myriad ways that occur in a cell as it propagates organization of process. All this is right in front of us, yet we can hardly say it.

Consider two molecules, A and B, that can undergo three possible reactions. A and B can form C and D; A and B can form E, and A and B can form F and G. Each of these three reactions follows what chemists call a "reaction coordinate." Each molecule has modes of motion: vibrations, rotation, and translation and complex combinations of these called librations. Different positions of the atoms in each molecule confer on that molecule a total energy with hills and valleys. A reaction from substrates to products tends to flow along valley floors, and over "saddle points" on "mountain passes" between adjacent valleys. These valley floors are the reaction coordinates of the reactions. The energy "hillsides" are internal constraints that tend to make the molecules remain in the valley floors. That is, they constrain the release of chemical energy in the reaction so that it flows in specific ways and forms specific products.

Now real cells do thermodynamic work to build lipid molecules, which then fall to a low-energy state of a bilipid layer, either in the cell membrane or in its internal membranes. Suppose A and B diffuse from the watery interior of the cell into such a membrane. The rotational, vibrational, and translational motions of these molecules will change. As a result, the energy hills and valleys will also change, thereby changing the very reaction coordinates and the constraints on the release of energy into work and the distribution of A and B into the different products of the three reactions. For example, in the aqueous phase, perhaps A and B exclusively form C and D. But in the membrane, the valleys change, and A and B form equal mixtures of E and of F and G. Thus the cell has carried out thermodynamic work to construct the lipids that formed the membrane, thereby altering the chemical energy constraints on the release of energy in A and B to make different reactions occur. The cells have done work to construct constraints such that energy is released in different ways. The cell has constructed constraints that alter the work that is done.

Obviously, the difference between forming C and D and forming E, F, and G can have propagating consequences of its own, and not only for the reactions that occur in the cell, which we know how to think about from chemical kinetic equations. What we do not yet have is a mathematics in place to study the propagation of construction of new constraints or boundary conditions on the release of energy. Ultimately we should join this mathematics with that of the familiar chemical, electrochemical, and mechanical processes to create a single coherent mathematical framework to describe the behavior of this cellular system.

Now notice that this propagating organization of process, in which subsets of the causal properties of these processes are "tasks" or functions relevant to the life of the cell (presumably assembled by natural selection), close upon themselves in a kind of "task closure" such that the cell constructs a copy of itself. This closure is a new state of matter—the living state. We do not understand it. While we have an intuitive sense of what this "task closure" means, at present we do not know how to describe it precisely. Clearly, the propagating process problem is not insoluble, but we have not integrated it into our understanding of the life of a cell, hence of life and its evolution in the biosphere.

This interwoven web of work and constraint construction, as well as the construction of other aspects of a cell, is simply awesome in its wonderful skein of persistent becoming. It has come into being by the natural process of the evolution of the biosphere with no need for a Creator God. It is at least part of what Kant was describing in his *Critique of Judgment.* A living cell is much more than mere molecular replication. It is a closure of work tasks that propagates its own organization of processes. The extent of this continuous propagation is wonderfully captured by biophysicist Harold Morowitz, who notes that the core of metabolism itself, called the citric acid cycle, is older than the oldest known rocks present on Earth today. The self-propagating organization of which Kant spoke is billions of years old.

This propagating organization of process is not deducible from physics, even though the "stuff" of the cell is physical and no physical laws are violated. Rather, the actual specific evolution in this specific biosphere, largely by natural selection acting on heritable variant, of the web of work and constraint is partly a result of Darwinian preadaptions. Nor is the evolution of this specific biosphere open to simulation by, say, the standard model in quantum mechanics and general relativity. We are very far beyond reductionism and into a universe of life, agency, meaning, value, and the still-to-be clarified concept of propagating organization of process.

What of the much-heard concept of information? Biologists are deeply fond of talking about information-rich molecules, and biology as an information processing science. It may be that no single concept of information can serve all of biology's requirements. This is a vexed subject that I shall dwell on only briefly.

One concept of information is Claude Shannon's famous example involving a source, a channel, and a receiver. Shannon, who invented his theory with respect to telecommunication channels, was mainly interested in "how much" information could be sent down a channel (such as a telephone wire) —in particular a "noisy" channel. He considered a source with a set of messages, one or more of which is sent down the channel to a receiver. Information sent down the channel and absorbed by the receiver reduces the uncertainty of the receiver with respect to the message sent down the noisy channel. This reduces the entropy about the messages at the source. It is important that Shannon never tells us what information *is*. He is only concerned with how much of it is moved down the channel. The "meaning" of the message is left out of Shannon's account.

Shannon leaves such "semiotic" problems of the meaning of the information, hence what information "is," to the receiver. But a bacterium is just such a receiver, and it responds to or interprets semiotic information—signs of food or poison by swimming towards or away from the sign, as appropriate. Thus the puzzling step to semiotic information, entirely missing from Shannon's information theory, requires of the receiver that it be, or communicate with, *an agent* (you, me, or a bacterium) that can potentially use the message to make things happen in the world (for example, the bacterium interpreting the sign by swimming up the glucose gradient). This step is a still inadequate effort to link matter, energy and information, which no one knows how to do.

The problem with applying Shannon's information theory itself to biology and the evolution of the biosphere is that we cannot make sense of the source, channel, or receiver. How does the transmission of information result in a human being? Is the source the origin of life? Is the channel the formation of offspring by two parents, or one parent via the fertilized egg, where the fertilized egg is the channel? There is not enough information storage capacity in all the molecular orientations in the fertilized egg to describe the resulting adult, so the fertilized egg cannot be the channel. If anything, development from a fertilized egg is closer to an algorithm than an information channel. In short, it seems that Shannon's concept of information is ill formulated for understanding the evolution of the biosphere.

Another concept of information was developed by the mathematician Andrei Kolmogorov, who explains that the information in a symbol string is equal to the shortest program that will produce it on a "universal computer," such as the one I am writing on. In other words, the information content of a sequence of numbers is defined by the algorithm needed to produce it. This has the surprising consequence that a string of random numbers has higher information content than any string of numbers with any pattern. More importantly, Kolmogorov, like Shannon, tells us how *much* information there is but not what information itself *is*. Kolmogorov's information is entirely devoid of "meaning" or semantic content.

In biology, the Shannon concept of information has focused on the famous genetic code and the coding by which a gene specifies an RNA which is translated into a protein. Some authors, such as John Maynard Smith, one of the best evolutionary biologists of the twentieth century, argue that the selection that genes have undergone picks genes out as "special information carriers of genetic information." Yet we can envision life-forms without genes (for example, collectively autocatalytic sets of peptides) or life as it originated before it evolved to have DNA, RNA, coding, and all the complexity of translation from RNA into proteins. A concept of information, if it is to apply to biology, should apply to any biology. Nothing fundamental seems to distinguish evolved DNA sequences, as "carriers of information," from evolved autocatalytic sets of peptides. What would we say of evolved autocatalytic sets of peptides in a lipid bounding membrane, but without DNA or RNA? They clearly have evolved. If they do work cycles like real cells, construct constraints, and create propagating processes of organization, aren't they alive—and therefore fulfilling their information requirements as fully as eukaryotic cells? What should we mean by the term *information*? And does it link to matter and energy, and if so, how?

My attempts in the last chapter and, briefly, above gave a Piercian account of what information "is": the discrimination of a sign, say, of a local glucose gradient, and interpreting that sign by an action, say, swimming up the glucose gradient. This is one set of thoughts about what information "is," and the start of how to link matter, energy, and information. If this view is correct, information requires an *agent, a nonequilibrium self-reproducing system doing work cycles*, to *receive* the information, *discriminate* it, and *interpret and act* on it. In living cells it is natural selection that has assembled the machinery to carry all this out, and natural selection has done so, presumably, for good selective reasons. Not all events in cells are selected, so not all events in cells are information laden. In this account, we begin to see one line of hints of how matter, energy, and information may come together. I should stress that this direction of thought differs from the Shannon and Kolmogorov interpretations of the amount of information, which, again, never say what information "is." I note again that for Shannon, whatever information "is," is left to the receiver. I think Shannon's receiver must ultimately be an *agent*.

But there may be a more primitive and profound sense of what information "is" than a semiotic sense with signs and interpretation of signs. In *What Is Life*, Erwin Schrödinger says that life depends upon quantum mechanics and the stability of chemical bonds. Then he says a periodic solid, like a crystal, is dull. Once you know the basic crystal unit you know the entire crystal, short of defects. A crystal of quartz is just a three-dimensional repetition of the unit crystal of quartz, with, again, some defects. Beyond the unit crystal and the defects, in a sense the crystal cannot "say" a lot. Schrödinger then brilliantly puts his bet on "aperiodic crystals," which he says will contain a "microcode" determining the resulting organism. His intuition is that an aperiodic crystal can "say a lot." He wrote this in 1943, ten years before the discovery of the structure of DNA—which is just such an aperiodic solid, with the famous arbitrary sequences of A, T, C, and G carrying genetic "information." His work stimulated the invention of molecular biology, mostly by physicists.

What did Schrödinger mean by "microcode"? My interpretation is that the microcode is a highly heterogeneous set of *microconstraints* that are *partially causal* in the myriad organized events that are unleashed in the cell and organism in its propagating organization of processes. In this sense, information is nothing but the constraints themselves. This interpretation has the merit that it unifies information, matter, and energy into one framework, for constraints are also boundary conditions. And physicists since Newton have thought of boundary conditions as partially causal with respect to the way the energy-laden physical system behaves. Following Maynard Smith, we can say that it is natural selection that has assembled these constraints, this information, because they are part of the propagating organization in cells and organisms that confers selective advantage. Unlike Maynard Smith (and most others who have thought about information in biology), I do not want to privilege genes as having information and the rest of the cell's intricate web of constraint related specific tasks as not having information. All have survived natural selection because they have proven useful. If we include semiotic information—in the form of the minimal molecular autonomous agent receiving a "yum" molecule bound to its "yum" receptor, hence a sign of "yum," and the agent acting to move toward the "yum," thereby interpreting the sign via an organization of processes achieved by natural selection—we may again be able to link matter, energy, and information into a unified framework.

We may then ask whether and how such ideas may generalize to the abiotic universe, where propagating organization of processes may also arise. Think of the cold, giant molecular clouds in galaxies, birthplaces of stars. Complex chemistry occurs; grains of silicates and carbonaceous material form, grow, and eventually assemble into planetesimals. Starlight and other radiation fields help drive the chemistry. The grains presumably have fractures on their surfaces, which serve as catalytic constraints that allow specific molecular reactions to occur, and some of the products may adhere to the grains, thus modifying their fracture surfaces, hence their boundary conditions, and thereby the further reactions that can occur on the grain as it grows. It seems worth asking whether this is an abiotic example of the organization of processes propagating. One can begin to see here how to unite matter, energy, and information.

Suppose we take, as a measure of the amount of "information" in a nonequilibrium thermodynamic system such as a cell, the diversity of constraints that are partially causal in the diversity of events or processes that happen next. Then we might hope that in living systems under natural selection the diversity of events or processes that happen next would be maximized. This seems a sensible intuition: cells that can carry out a wide diversity of processes seem likely to outcompete those that cannot. But this hope raises an immediate problem. I ask you to be a physicist for a moment. Recall that we can describe the positions and velocities of any particle in a three-dimensional space by six numbers, three for the projections of the position onto a chosen X, Y, Z axis coordinate system and three to give the velocities, including their X, Y, and Z directions and magnitudes. Then $6N$ numbers describe N particles in what we called a $6N$-dimensional state or phase space. Now as an amateur physicist, think of a gas of randomly moving particles. The diversity of what can happen next is maximized, for the N particles might move *in any possible combination* of directions and velocities. Crucially, *such random motion of the gas particles is pure heat.* Yet, and this is critical, there are precisely *no* constraints in this *random* system. Thus, the diversity of what can happen next is maximized in a system of particles in random motion, not in a system with boundary conditions or constraints. This maximization of what can happen next in the fully random motion of particles is not what we seek in trying to articulate how DNA has a *microcode* for the very nonrandom generation of the adult, and has parallels to the mathematical fact that Shannon information is maximized in a random string of symbols. Random motion of particles, as noted however, constitutes pure *heat*, not *work*. This may provide the clue we need to find a useful conception of a sense in which a diversity of constraints maximizes the diversity of "events" that can happen. If constraints and their diversity are somehow to be related to a physical meaning of information and to "events," then we need to introduce boundary conditions on the maximally random motion of the random gas—to have fewer motions, and thus constrain the release of energy into fewer degrees of freedom. But the release of energy into few degrees of freedom constitutes *work*. *So I suspect we need a concept of information as constraints on the release of energy that then constitutes work, and then hope to show that natural selection maximizes the diversity of work that is done in cells, organisms, ecosystems, and biospheres.* "Events," then, become coordinated, or correlated motions of particles due to the constrained release of energy, thus work done. This seems promising. But mere diversity of work also seems inadequate to what we might suppose selection has maximized. We also need a notion of the total *amount* of work done. The total amount of work a system can do depends upon the energy flow through the system. It would seem that we need to consider the product of the total *amount* of work done multiplied by the *diversity* of work that is done. This is mathematizable. For a cell, this would maximize the total amount of work times the diversity of work in the selectable tasks the cell could carry out. Therefore, it seems deeply important that ecologist R. Ulanowitz has introduced a similar mathematical measure for work flows in ecosystems. The measure is the total energy flow in an ecosystem times the diversity of that energy flow. He has shown that mature ecosystems maximize this measure, while it drops to lower values for perturbed ecosystems. Energy flow and work flow are nearly the same thing. Work ignores the pure heat present in energy. If the reasoning above and Ulanowitz are right, life, due to natural selection, within cells, in ecosystems and the biosphere, may maximize something like the product of total work times the diversity of work done

In the next chapter, I discuss old and recent work that raises the possibility that cells may be dynamically "critical," poised between dynamical order and chaos, thus at the "edge of chaos," as defined in that chapter. *Critical dynamical systems maximize the correlated behavior of variables in systems of many variables. Also critical dynamical systems appear to maximize the diversity of what they can "do" as they become larger. This raises the fascinating but unproven possibility that, due to natural selection, life achieves a maximization of the product of total work done multiplied by the diversity of work done by being dynamically critical. Then cells would be maximally efficient in carrying out the widest variety of tasks with the maximum total work accomplished, given energy resources available.* As we will see, there is early, tentative evidence that cells may actually be dynamically critical.

Meanwhile, very early work is underway to try to establish a connection between dynamical criticality and thereby the simultaneous maximization of total work times the diversity of work done in cells, and perhaps ecosystems and biospheres. If this were true for biospheres, then, despite extinction events, their long term behavior would be to maximize something like the total diversity of organized processes that can happen.

Something like this just might be a law for self-consistently self-constructing, open, nonequilibrium systems such as cells or biospheres anywhere in the cosmos. Biospheres may maximize Kant's self-propagating organization of process.

All of this evolution of propagating organization of process in and among cells, linking matter, energy, work, constraint, and semiosis, was going on in the biosphere billions of years ago, before the evolution of multicelled organisms, and is part of the evolution of the biosphere into its adjacent possible. We are the fruits of this biosphere. We can only have profound gratitude to participate in this ongoing evolution. The creativity in nature should truly be God enough for us.

8

ORDER FOR FREE

Self-organization that is both emergent and not reducible to physics is not limited to some theories of the origin of life. It is also visible in your development from a fertilized egg through the process of ontogeny. This emergent organization is part of the propagating organization of process discussed in the last chapter. However, so far, what I shall describe in the present chapter about the dynamical behavior of cells has not been united with the concepts of propagating organization of process, work, constraint, and the hoped-for possibility that cells maximize the product of total work done times the diversity of work done.

Self-organization is very well established as a mathematical fact, and there is early evidence that the same self-organization actually applies in the bodies of plants and animals. Whether this self-organization actually applies to contemporary cells or not, it is a clear example of the kind of spontaneous order in biology that might occur. It is, at a minimum, a concrete example of the *possibility* of emergence, nonreducibility, and, as we shall see, the powerful idea that order in biology does not come from natural selection alone but from a poorly understood marriage of self-organization and selection. Thus the classical belief of most biologists that the only source of order in biology is natural selection may well be wrong. Self-organization, a second source of order, lies at hand as well, to mingle in yet unknown ways with natural selection and historical frozen accidents.

Ontogeny is a magically complex process of emergent order. A human starts as a single cell, the fertilized cell or zygote. This zygote undergoes about fifty cell divisions and creates all the different cell types in a newborn baby: liver, kidney, red blood cells, muscle cells—about 265 cell types by histological criteria. This process of creating different cell types is called "differentiation." Roughly speaking, there is a branching tree of differentiation from the zygote to all the final cell types of the adult organism. Some cells remain as "stem cells" capable of giving rise to specific subtrees. For instance, the blood stem cell gives rise to all the cell types of blood in an adult. The differentiation tree also has a few cross-connections among different branches. In addition to differentiation, the diverse cell types give rise to specific organs (liver, arm, eye) in a process called morphogenesis. While I focus here on self organization with respect to differentiation, there are signs of self organization in morphogenesis as well.

Here is the central dogma of molecular biology. Most of us are familiar with DNA, whose double-helical structure was discovered by Watson and Crick in 1953. DNA comprises an alphabet of four chemical letters, A, C, T, and G. These are arranged in complex sequences along each strand of the double helix, with A on one strand bonded to T on the complementary strand, while C is bonded to G. A sequence of three DNA letters, such as CGG, called a coding triplet, codes for one of twenty kinds of small, organic molecules called amino acids. In a process called transcription, one part of one DNA strand is copied into a complementary strand of a similar molecule called RNA. (In RNA the letter T is replaced with a U, but that is a minor detail.) In a process called translation, a molecule called a ribosome binds to the RNA at a translation "start site," reads each RNA coding triplet, and links the specific amino acid for which that triplet codes onto a growing chain of amino acids called a protein. Thus our genes are mainly sets of instructions for building proteins.

Once assembled, the linear protein becomes functional by folding into a complex three-dimensional shape. Proteins are the chemical "workhorses" of the cells, building specific structures, making cell membrane channels, and working as enzymes catalyzing specific reactions.

A DNA sequence that encodes a protein is a structural gene. The Human Genome Project has shown that humans have about twenty-five thousand different structural genes. But because of some flexibility in translation, the number of distinct proteins in human cells may be far higher, perhaps a hundred thousand or more.

The DNA in a cell is folded into a complex structure called a chromosome, and, within the chromosome, the folded DNA and ancillary proteins attached to it are called chromatin. Most of us have twenty-three pairs of chromosomes, each pair consisting of one chromosome from the mother and one from the father.

A central question in biology is how cell types come to be different from one another. In the early twentieth century, it was thought that while the zygote contains all twenty-three pairs of chromosomes, perhaps different cell types would contain different subsets of these chromosomes. Microscopic examination soon showed that all cell types contain the full set. Then it was discovered that if one dissociates a carrot into all its different cell types and cultures them appropriately, each cell type can give rise to the whole carrot. This means that each cell type contains the genetic information to make the entire set of cells types in the full carrot, so different cell types do not arise from a loss of genetic information. Shortly after, an embryologist named Hans Driesch showed that if a frog embryo at the two-cell stage is separated, each cell gives rise to an entire normal frog. Driesch was so stunned by this phenomenon that he developed a theory of entelechies, an obscure kind of ordering principle, to account for the facts. No one takes these ideas seriously today, but we all remained fascinated by his observations on the frog embryo.

The next step in the understanding of cell differentiation came with the demonstration that different cell types make different sets of proteins. For example, red blood cells make hemoglobin. Certain white blood cells make antibody proteins. Muscle cells make actin and myosin. The presence of these proteins means that different genes are active in the different cell types.

The fundamental breakthrough in thinking about cell differentiation came from the brilliant work by two French microbiologists, François

Jacob and Jacques Monod, in 1961 and 1963. They demonstrated that some proteins bind to the DNA very near specific structural genes encoding some proteins, at DNA sequence sites variously called operators, promoters, enhancers, and boxes, and by binding tend to turn the nearby structural gene on or off. That is, these regulatory proteins, often called transcription factors, increase or decrease the transcriptional activity of various genes. In turn, this increases or decreases the RNA made from that gene, hence increases or decreases the protein encoded by that RNA. In simple terms, Jacob and Monod discovered that genes turn one another on and off.

In their famous 1963 paper, Jacob and Monod speculated how regulatory genes could account for cell differentiation. Imagine, they said, two genes, A and B, each of which is spontaneously active but turns the other gene off. This little genetic circuit has a deeply important property. It has two different stable patterns of gene activation: A on and B off, or A off and B on. Thus this little genetic circuit—then entirely hypothetical—could contain the same set of genes but support two different cell types.

This prediction was demonstrated through experiments by Neubauer and Calef in the early 1970s. They noted that a virus that infects bacteria, called bacteriophage lambda, has two genes, *C1* and *Cro*, that turn each other off. That is, each gene represses the other. In certain mutant lambda-infected bacteria, this little circuit could persist in either the "*C1* on, *Cro* off" state or the "*C1* off, *Cro* on" state for hundreds of cell divisions. Occasionally the circuit would jump from one pattern to the other. Interestingly, human cells also occasionally "jump change" from one type to another in odd ways called metaplasia: pancreatic cells change to liver cells, uterus cells to ovarian cells, and so forth.

Thus, genes in cells form a complex transcriptional regulatory network in which some regulatory genes enhance or inhibit the activity of themselves or other genes. Often many transcription factors bind to complex promoters, and the activity of the regulated gene is a complex function of the combination of regulating factors. Budding yeast is a well-studied, complex regulatory network with about 6,500 structural genes and about 350 different transcription factors. The regulatory network is one giant interconnected web, at least the half of it that is known. The 350 regulating genes, perhaps some other genes that code for what are called transcriptional cofactors and cell signaling proteins, and perhaps other molecules regulate one another's activities and also regulate the roughly 6,500 structural genes.

In humans, it is estimated that about 2,500 regulating genes encode 2,500 different transcription factors and cofactors. These 2,500 genes and other signaling proteins regulate one another and the remaining 22,500 genes. Recently, microRNA has been discovered. These genes encode short RNA molecules that bind the RNA transcribed from other genes and destroy them, shutting off the translation of that RNA into its protein. MicroRNA is also part of the genetic regulatory network, as are molecules that modify chromatin structure. Jacob and Monod's little two-gene mutually inhibitory circuit is, in real life, a vast web of mutually regulating interactions among 2,500 or more genes and their products, and controlling 22,500 or so other genes. If the two alternative steady states of the Jacob-Monod circuit correspond to two hypothetical cell types, what massively complex configuration of on and off corresponds to our own cell types?

RANDOM BOOLEAN NETWORK MODELS OF GENETIC REGULATORY NETWORKS

One approach that allows us to think about cell differentiation in humans and other multicelled organisms is to make overly simple mathematical models of possible genetic networks and study their behavior. This work has been underway for over forty years.

Real genes are not just "active" and "inactive," they can show graded levels of transcriptional and translational activity. But if we want to study networks with thousands of genes, it is helpful to begin by idealizing a gene as if it were as simple as a light bulb. The idealization is false, but it's useful. After all, physicists have been using hard elastic sphere models of atoms in statistical mechanics for over a hundred years with considerable success, even though we know that atoms are nothing like hard elastic spheres.

When I started my own work on this topic in 1964, no one had any idea of the structure of real cellular genetic regulatory networks. Now we do, but at the time, I began with simple questions: Does the genetic regulatory network controlling ontogeny have to be specially constructed by natural selection to accomplish its miraculous feats? Or might even random networks behave in ways that were close enough to what we know of biology that their spontaneous order ("order for free," I call it) might be available for selection's finer siftings?

Astonishingly, "order for free," much like real biology, does exist in random-model genetic light-bulb networks. This is emergent self-organization, not reducible to physics. The name for such a network is random Boolean network, because the rules governing the on/off behavior of any regulated gene as its regulatory inputs vary from off to on is given by a logical or Boolean function.

Figure 8.1a shows a small Boolean network with three on/off, or binary, genes, A, B, and C. Each gene is regulated by the other two. These regulatory connections can be thought of as the wiring diagram of the genetic network. Each gene has four possible present states of its two inputs: 00, 01, 10, and 11. The first digit in each pair refers to the first input, and the second digit refers to the second input. (So 01, for instance, means first input off, second input on.) Having made the wiring diagram, we need to specify how each gene will respond. The response of the regulated gene is given by a logical, or Boolean, function that specifies for each of the four input states the output activity, 1 or 0, of the regulated gene. In figure 8.1a, gene A, regulated by B and C, is governed by the logical "and" function: A is active at the next moment only if both B and C are active at the present moment. B and C are each governed by the Boolean "or" function, meaning that the regulated gene is active at the next moment if either input, or both, are active at the present moment.

In the simplest model, we make two further assumptions. First, time comes in *discrete* clocked moments, 0, 1, 2, etc. At each clocked moment, all genes *simultaneously* consult the activities of all their inputs, consult their Boolean function, and assume the next state of activity, 1 or 0, as specified by their Boolean function for the given input activity pattern.

Both assumptions are major idealizations. Real genetic networks have neither clocks nor synchrony. We will remove these idealizations later.

Figure 8.1b shows all possible combinations of the simultaneous activities of the three genes, A, B, and C, at the current moment T, and the successive activities of each gene at the next clocked moment T + 1. A *state* of the network is defined as any of the possible combinations of simultaneous activity of the three genes. Since each has two states, 0 and 1, there are 2 x 2 x 2—or eight— possible states of this little network. These eight states constitute what is called the "state space" of the network, namely all the states the network can be in.

In figure 8.1b, the columns on the T + 1 side of the figure below each gene show its responses at time T + 1 to the activities of its inputs at time T. These columns are just the Boolean functions from figure 8.1a, rewritten. The interesting thing is that the rows on the T + 1 side of the figure show the network states to which each network state on the T side of the figure "flows" one clocked moment later. Thus the state (000) "changes" to the state (000) which happens to be the same state! Thus (000) is a steady state of this little network. If the network is released in the state (000) it will remain there forever unless perturbed in some way.

Nor consider a Boolean network with N genes. Just as there are 2^3 states for the 3 gene network, there are 2^N states for a network of N genes. Before we know anything else, this is already important. Humans have about 2,500 transcription-factor genes. The number of possible combinations of activities of these genes is $2^{2,500}$ or about 10^{750}. There are only an estimated 10^{80} particles in the known universe, which means that the number of possible states of our regulating genes is enormously, vastly larger than the number of particles in the universe. If we add in our structural genes, there are $2^{25,000}$ or about $10^{7,500}$ states. The universe is about 10^{17} seconds old. It takes a few minutes to an hour for genes to turn on or off. Even if we were a thousand times as old as the universe, there is not time enough for our regulatory network to "visit" even a tiny fraction of its possible states. And remember, we are still assuming genes are as simple as light bulbs, with no states but "on" or "off."

Furthermore, cell types are clearly distinct from one another and stable, often for long periods. Thus each cell type must somehow represent a very confined set of these unimaginably many possible states. And these confined patterns must somehow be clearly distinct from one another.

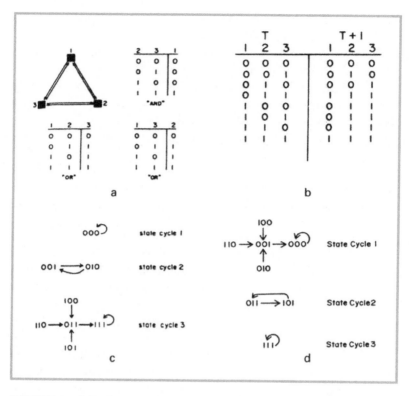

a

b

c

d

FIGURE 8.1 A, B, C, and D   (a) The wiring diagram in a Boolean network with three binary elements, each an input to the other two. (b) The Boolean rules of (a) rewritten to show for all 2 raised to the 3rd = 8 states at time T, the activities assumed by each element at the next time moment, T + 1. A "state" of the network is any combination of the activities, 1 or 0, of the three elements. All elements "update" their activities simultaneously at time T + 1. Read from left to right, this figure shows the successor state for each state. (c) The state transition graph, or "behavior field" of the autonomous Boolean network of (a) and (b), where "autonomous" means that there are no inputs to the network from outside the network. The behavior field, also called the "dynamical flow in state space," is obtained by showing the state transitions to successor states connected by arrows. (d). A mutation of the initial network in figure 8(a), in which the Boolean rule of gene 2 is changed from "OR" to "AND," similar to gene 1. The result changes the behavior field, creating a novel state cycle 2, leaving the two steady state cycles of (c), 000 and 111, but shifting the transients so now state 000 drains a large basin of attraction along transients that flow into it.

How might such confinement arise? Consider a network of N genes released from an initial state. At each clocked moment, the network will pass to some successor state along what is called a "trajectory" in state space. But there are a finite number of such states, 2^N. Eventually the network will reenter a state it has visited before. Since the system is deterministic, it will do the same thing it did the last time, hence enter into some kind of closed cycle of states around which it "orbits" forever if not perturbed. The closed cycle of states is called a state cycle. The state cycle is also called an attractor, for the important reason that many trajectories often to lead to the same state cycle. The set of states lying on trajectories that end on one specific state cycle, plus that state cycle itself, are called the attractor's basin of attraction. An attractor is rather like a mountain lake, collecting flow from many streams in its drainage basin. Thus the natural way to confine the network's behavior to subregions of its state space is simply to let it evolve through state space under its own dynamics. Confinement arises inevitably as the system flows along a trajectory to an attractor and then, if unperturbed, becomes trapped there forever.

Inevitably, biologists who are interested in such things formed the hypothesis that a cell type corresponds to an attractor, not to any of the transient states flowing to the attractor. Recent evidence is beginning to confirm this hypothesis.

But humans have some 265 cell types. Can there be that many attractors in our genes' state space? Happily, random Boolean networks typically have multiple attractors, each draining its own basin of attraction, like many lakes in a mountainous region, each draining its own catchment basin.

Figure 8.1c shows these attractors for the network in figure 8.1.a. As you can see, there are three attractors. Two are steady state patterns of A, B, and C gene activities, (000) and (111), and the third is a little oscillation (001)-(010)-(001). The (111) attractor drains a modest basin of attraction.

ORDER FOR FREE

Even random Boolean networks can exhibit staggering self organized and emergent order. I discovered this as a medical student in 1965. I programmed a N = 100 gene network, with each gene receiving K = 2 randomly chosen inputs from among the 100, and each assigned responses at random from the possible 16 Boolean functions with K = 2 inputs. Such a network has 2^{100} states, or about 10^{30}. I was paying for the computer time myself. If the state cycles were going to be long, I was going to go broke.

To my delight, the very first network had only five state cycles, four that cycled through four states, and one that oscillated between only two states. The system, with 10^{30} states, one with thirty zeros after it, *spontaneously confined itself* to just five attractors, with either four or two states. If my state space were real space, these attractors would have been infinitesimal black holes within its vast reaches.

I did a lot of work investigating random Boolean networks of various sizes. It turned out from numerical evidence that the median number of states on a state cycle was the square root of N. (Now this is an established mathematical theorem.) As N grows, the state space grows as 2^N, hence exponentially, and the square root of N grows more slowly than N itself. In big networks, typical state cycles are miniscule. Self-organization that confines patterns of model gene activities to tiny regions of the network's state space arises spontaneously in these networks. There is order for free.

Beyond the enormous confinement obtained, these results are critical to the time scale of cells. As noted, it takes a few minutes to an hour to turn a gene on or off. Thus, with 25,000 total genes, such a network would take about 160 to 1,600 minutes to orbit its state-cycle attractor. This is exactly in the biological ballpark—all without building anything but random K = 2 networks.

Below I will show that these early results generalize to a much richer class of "critical" Boolean networks, where K = 2 networks are also critical, and show early evidence that cells are dynamically "critical."

But if you had such a random N = 10,000, K = 2 network, the wiring diagram among its 10,000 genes would be a mad scramble of wires, and the logic of the genes is utterly random. Yet this order arises spontaneously.

Cell types are clearly distinct from one another. A powerful feature of the dynamical network models we are describing, whether random Boolean networks, or, later, more refined models, is that the distinct attractors of the same network can correspond to distinctly different cell types with distinctly different patterns of gene activities. We account naturally, therefore, for the observed fact that cell types are distinct.

What about the number of lakelike attractors? I suggested, from numerical evidence, that these too scaled as the square root of N. More recent work shows that this was wrong. I undersampled very large state spaces, and the number of attractors increases faster than N. But even more recent work shows that if the synchrony assumption is dropped—if you don't require every gene to respond in unison, at every step—the median number of stable attractors increases as the square root of N, the number of genes in the model.

This was rather astonishing. *Here was a theory that had a hope of predicting the number of cell types as a function of the number of genes.* It should be roughly a square root function. Still a medical student, I graphed the number of known cell types in simple and complex organisms from bacteria to humans and plants, and found that, in fact, the number of cell types scaled as the square root of total DNA per cell! Of course I felt excited. But I did so with a caveat. Even in 1965 it was thought that not all DNA consists of structural genes. Much of it came to be called "junk" DNA, its function unknown. It is not clear that the number of structural genes scales in direct proportion to the DNA per cell. If it does, my prediction would be fine. The number of regulatory genes, however, includes the newly discovered microRNA. The number of microRNA genes is still unknown. Using estimates of the number of regular structural genes in different organisms, the number of cell types appears to increase at most linearly with the number of genes. That is, a doubling of the number of genes roughly doubles the number of cell types. All this will be refined as we refine our molecular understanding of what counts as a distinct cell type. Presumably, with this refinement, we will distinguish more cell types. It remains to be seen how the number of distinct cell types scales with the number of genes.

Whatever the final answer may be, random Boolean networks limited to two inputs per gene (a constraint not present in real networks) predict that organisms should have only a modest number of discretely different attractor cell types. Again, this is very much like real biology with, as yet, no selection—but read on.

Dynamical Order, Chaos, and Criticality

I can now summarize over forty years of work on random Boolean networks, with fixed numbers of inputs per gene and various complex input distributions more closely matching what is beginning to be known of the genetic networks in real cells.

Briefly, these networks exhibit three regimes of behavior: ordered, chaotic, and critical, i.e., posed at the boundary or edge between order and chaos.

In the ordered regime, state cycles are tiny. In the critical regime, state-cycle lengths scale as the square root of N in the known cases. In the chaotic regime, state cycles are vastly long, having a length that is exponential in N. In the limit where $K = N$ (meaning the number of inputs equals the number of genes—every gene is connected to every other), random Boolean networks have state cycles on the order of $2^{N/2}$. A network with two hundred genes would have a state cycle with 2^{100} states, and take billions of lifetimes of the universe to complete one orbit at a microsecond per state transition. Chaos is not order for free. I don't want my liver cells doing that, nor do you.

In the ordered regime, a large fraction of genes become frozen, constantly on or constantly off. These form a connected frozen subnetwork that *percolates* across the network and scales with the size of the system, N. This frozen sea leaves behind isolated islands of genes that fluctuate in activities or show alternative steady state activities on different attractors. Each such island is *functionally isolated* from the others by the frozen sea, and therefore the set of all the attractor model cell types of the entire network is just the set of all combinations of attractors of all of the island. Thus, suppose there were, say, three attractors for the first functionally isolated island, five for the second island and two attractors for a final third island. Then the entire network would have 3 x 5 x 2 = 30 total cell type attractors. This partitioning of the network into functionally isolated islands in the ordered—and critical—regimes, each with its own alternative attractors, is interesting. The alternative attractors of one island can be thought of as a genetic "memory switch" able to "remember" what attractor it is in. Then the behavior of the entire genome consists in a kind of combinatorial "epigenetic" code, in which each cell type is specified by the states of its distinct genetic memory switches.

In the chaotic regime, a twinkling sea of genes following tremendously long state cycles spreads across the network, leaving behind frozen islands.

As we tune parameters such as K (the number of inputs per gene) or the choice of Boolean functions, we can shift networks from the chaotic to the ordered regime. During this shift the networks pass through a phase transition where they are dynamically critical. Here, the fluctuating sea is just breaking up into isolated fluctuating islands.

The response of the network to transient perturbations, "flipping" the activity of a single gene, differs markedly in the ordered, critical, and chaotic regimes. We can define a gene as dynamically damaged if, following a perturbation to some other gene, the gene in question ever behaves (even once) differently than it would have without the perturbation. Once a gene is damaged it remains so, even if its behavior returns to normal. With this definition we can see how transient perturbations cause damage "avalanches" and, more importantly, we can ask how big such damage avalanches are. In the chaotic regime, there are many vast avalanches in which perturbation of a single gene alters the activities of 30 percent to 50 percent of the genes. This would correspond to altering a single human gene and seeing ten thousand or more genes alter their activity. This *never* happens. The typical damage avalanche in real human cells is a few tens to a few hundreds of genes. This suggests that real cells are not chaotic.

In the ordered regime, damage avalanches are small and fall off in frequency exponentially with avalanche size. In other words, there are some large avalanches but they are very rare.

In critical networks, we see a very special distribution of avalanche sizes called a power law. Measure a very large number of avalanches. Plot on the X axis the size of the avalanche, and on the Y axis the number of avalanches of that size. The result is a histogram of the distribution of avalanche sizes. Now, transform the X and Y axes to the *logarithm* of the size of the avalanche and the *logarithm* of the number of avalanches of that size. In this "log-log" plot, the avalanche size distribution for critical networks is a straight line angled down to the right, and the slope of the

angle is −1.5. This straight line in a log plot is called a power-law distribution. As discussed below, just such an avalanche distribution has been
found in yeast cells—early evidence that real genetic regulatory networks
may in fact be critical.

The next distinguishing feature of ordered, critical, and chaotic networks is whether nearby states lie on trajectories that converge (as in ordered networks), stay the same distance apart (critical networks), or diverge
(chaotic networks). The divergence, or spreading apart of trajectories, in
chaotic networks is the same thing as the vast damage avalanches, or sensitivity to initial conditions, seen in the chaotic regime. Nearby states, even
single gene flips, lie on trajectories that spread far apart. In the ordered
regime, nearby states converge and typically flow to the same successor
state along converging trajectories. This is why damage from perturbations
seldom creates large avalanches: a gene perturbed into a neighboring trajectory quickly returns to its former state cycle. In the critical regime, nearby
trajectories stay the same distance apart.

The Critical Subset of
Random Boolean Networks

At present we know three ways to "tune" random Boolean networks from
the ordered to the critical to the chaotic regime. First, we can increase the
mean number of inputs, K. At $K = 0$, networks are in the ordered regime.
If there are no inputs, nothing can change. At $K = 2$ networks are critical.
(By luck, I stumbled across critical networks on my first try.) For K greater
than 2, networks are chaotic.

A second means to tune from order to chaos occurs if K is greater than
2. We can bias the Boolean functions so that most input combinations
yield the same response, either a 1 or a 0. Such a bias is called a P-bias,
where P is the fraction of input combinations that yield the same response
by the regulated gene. In a parameter space of K and P, a thin curve corresponding to the phase transition separates the ordered regime from the
chaotic regime. Thus critical networks are very rare in the space of networks. If real cells are actually critical, this suggests that it must be very

important to be so and that this is a condition maintained by powerful natural selection.

A third way to tune from chaos to critical to order is to use a power law distribution of inputs to the gene, and tune the slope of the power law. It has been found that critical networks correspond to a power-law slope of between −2.0 and −2.5. Strikingly, many observed real-life networks have input distributions with power-law slopes in this range. Again, critical networks correspond to a thin line in this two-dimensional parameter space, hence are very rare. In very early published and unpublished results on the regulatory input distribution in human cells, the distribution in two large data sets, analyzed in two entirely different ways, appears to be a power law distribution with a slope of about −2.5. At this stage, this is at most intriguing, but may prove true with further work.

WHY IT MIGHT BE USEFUL
FOR CELLS TO BE CRITICAL

We do not yet know whether cells are critical. However, there are many reasons why it might be advantageous to be so. The first of these is that cells must coordinate past discriminations with reliable future actions. Suppose a genetic network, or more broadly a cellular causal event network propagating organization of process, is deep in the ordered regime. Suppose two distinct states correspond to discriminating food from poison. If the two states lie on trajectories that converge on the same successor state, the system has "forgotten" where it came from and lost the discrimination between food and poison.

Conversely, suppose that the network is chaotic. Small fluctuations will cause vast damage avalanches that yield very different future behavior. Then the system cannot act reliably in the presence of slight noise. Worse, there is always slight noise, meaning the system cannot ever act reliably.

Critical networks, poised between order and chaos, seem best able to coordinate past discriminations with reliable future actions. And, in fact, recent numerical results suggest that this is correct.

A second reason why criticality could be advantageous is a quite remarkable mathematical result. A measure called "mutual information" can be used to gauge the correlation between the fluctuating activity of two genes or events—one now, one a moment later. Mutual information, which is the sum of the entropy of each gene minus their joint entropy, is zero if the two genes are fluctuating randomly with respect to each other, or if either or both are not fluctuating. It is positive (with a maximum of 1.0) if the genes, or events, are fluctuating in a correlated way. My colleagues, A. Ribeiro, B. Samuelson, and J. Socolar, and I have recently shown that average mutual information between pairs of genes, A and B, one moment apart, is maximized for critical random Boolean networks. This suggests that the most complex coordinated behavior that can occur among random Boolean networks is in critical networks. Although it's unknown how this may bear on nonrandom networks, it is very interesting that maximum correlation among many pairs of variables does definitely happen for critical networks in the vast class of random Boolean networks. The intuition derived from the maximization of mutual information, or coordinated altering behaviors among pairs of variables, in critical networks is that the most complex coordinated behavior can occur in critical networks. This intuition squares with the fact that in critical networks the fluctuating chaotic sea is just breaking into isolated fluctuating islands, so again, the most complex, but organized behavior should occur in critical networks. In more ordered networks the behavior would be more "frozen" and less complex. In the chaotic networks, slight noise would dramatically alter network behavior. It would seem that the most complex computation could occur in critical networks in the presence of slight noise, which would disrupt chaotic networks. Certainly what happens here is more interesting than the dull behavior found deep in the ordered regime, where most genes are frozen in fixed states.

A final result by Ilya Shmulevich extends the intuition that the most complex coordinated behavior can arise in critical networks. Shmulevich created ordered, critical, and chaotic random Boolean networks. He defined the "weight" pi, of a basin of attraction, i, as the number of states in that basin of attraction, that is, the basin of attractor i, divided by the total number of states, 2 to the Nth power, in the state space. He then used a familiar mathematical formula, the entropy formula, to calculate the entropy of the network. This formula is (minus the sum of pi times log pi) where the sum is over all the basins of attraction. In effect, this entropy is a measure of the diversity of behaviors of the network's attractors. Shmulevich found that for ordered and chaotic networks, as the number of model genes, N, increases, the entropy increases for a while, then stops increasing! The behavior does *not* become more diverse as the system size increases. In contrast, in critical networks, the entropy, or diversity of behaviors, continues to increase as network size increases. Thus, only critical random Boolean networks can grow large and carry out increasingly diverse, complex, coordinated behaviors.

The existence of an ordered, critical, and chaotic regime is not limited to random Boolean networks. Leon Glass showed that a family of what are called piecewise linear genetic network models have the same three regimes. Work is currently underway using what are called chemical master equations, the most realistic models we have of the detailed molecular reactions in genetic regulatory networks. Early results show an ordered regime and a chaotic regime. Presumably these are separated by a critical phase transition, but this remains to be demonstrated.

TENTATIVE EVIDENCE
THAT CELLS ARE CRITICAL

Several lines of experimental and theoretical work are beginning to suggest that real cells, or their genetic regulatory networks, may be dynamically critical.

The first evidence concerns damage avalanches. Two groups, Ramo et. al. and Serra et. al., deduced that the deletion of a gene in a critical network should alter the activities of other genes with a power-law distribution whose slope is −1.5. It is quite incredible that analysis of 240 yeast mutants, each deleting a single gene, resulted not only in a power-law distribution of avalanches but that their slope is, as predicted, −1.5. Two points are of interest. First, the successful prediction, which delights me even if I would not yet bet my house that cells are critical. Second, this is a new biological observable—the size distribution of the avalanches of alterations in gene activities following deletion mutants is not a standard question in molecular biology. But it reflects an emergent collective property of critical networks and thus could prove extremely useful.

A second line of evidence is my work with a colleague, I. Shmulevich, that analyzes a time series of gene expression data for all twenty-five thousand genes for forty-eight hourly time points. A complex analysis is strongly inconsistent with these human cancer HeLa cells being chaotic, but consistent with HeLa cells being ordered or critical.

A third line of evidence, the "four kingdom" results, concerns bacteria, fruit flies, a plant called arabidopsis, and yeast. In all four cases, scientists have made very successful Boolean models that capture the effects of many normal and mutant genes. Analysis of these models shows that all are critical. This result is highly unexpected because critical networks are very rare. A selection pressure that causes such vastly different organisms all to be critical (remember, criticality represents an extremely tiny fraction of the entire state space) must be very powerful.

The final line of evidence measures whether nearby trajectories converge, diverge, or stay the same distance apart over time. The experimental data concern stimulation of particular cell surface receptors on blood cells, called macrophages. Gene activities were monitored for all twenty-five thousand genes over time, collecting five time moments a half hour apart, for six such perturbations. The result is over three hundred data points showing, stunningly, that *trajectories over time stay the same distance apart,* neither converging as in the ordered regime, nor diverging as in the chaotic regime. The data are consistent with criticality. Interestingly, a deleterious mutant of a white blood cell was the only one to show divergence, indicative of chaotic behavior. Thus the evidence for criticality in normal cells is not trivial at all.

All of these observations are new biological observables of collective emergent properties of genetic networks, not familiar yet to many biologists, but fully legitimate and a major new means to understand the integrated behavior of genetic regulatory networks. It seems very likely, but remains to be shown, that the simplest case of Boolean networks, and Glass's piecewise linear genetic network models, are telling us that an enormously wide class of nonlinear dynamical systems exhibits order, criticality, and chaos. More research is needed. We need to know if real genetic networks, or causal networks in cells, are actually critical.

These properties of order, chaos, and criticality are independent of the specific physics involved. They rest on mathematical features of complex networks. Like the collectively autocatalytic sets we saw in chapter 6, these emergent properties of gene networks are not reducible to physics. They are Laughlin-type organizational laws. The existence of multiple examples leads to the conclusion that emergence is not rare. It is all around us. We will see more of this in considering the economy as an emergent, largely unpredictable, system. We truly need a new worldview, well beyond the reductionism of Laplace and Weinberg. Finally, these results for vast classes of model genetic regulatory networks suggest that self organization, order for free, is as much a part of evolution and natural selection as historically frozen accidents. We must rethink evolution.

9

The Nonergodic Universe

I now wish to discuss what will turn out, among other things, to be the physical foundation for much of the evolution of diversity in the biosphere, of economic growth and even human history. The present chapter lays the foundations for the two that follow, where I will argue that the evolution of the biosphere and economy and history are all ceaselessly creative in ways that we typically cannot foretell. Thus, not only are we beyond reductionism into a scientific worldview that includes "emergence," we will see that we are coming to a scientific worldview that takes us into subjects normally thought to lie outside the realm of science.

At levels of complexity above atoms, the universe is on a pathway, or trajectory, that will never repeat. For example, in the evolution of the biosphere, from molecules to species, what arises is almost always unique in the history of the universe. Using the physicist's technical term, the evolution of molecules and species in the biosphere is vastly nonrepeating, or nonergodic.

Consider a small volume like our cylinder from chapter 7, with a few atoms of gas in it. Each atom has a position in the volume, and a velocity—a direction and speed of motion. In three-dimensional space, the position projected on three spatial axes can be specified by three numbers. The velocity projected onto the three axes can also be specified by three numbers. So six numbers characterize the position and velocity of each of the N particles of gas. If there are N such particles, $6N$ numbers specify the current configuration of the gas. Next, note that we know beforehand the volume the gas occupies. That is, we know the space of possible configurations of the gas, or equivalently the $6N$-dimensional phase space of the gas system.

Now divide the $6N$ phase space into many tiny boxes, called microstates. Over time, the particles collide and follow Newton's laws. The gas system flows in the $6N$-dimensional space along some trajectory. According to the famous (and perhaps false, based on computer studies several decades ago) ergodic hypothesis, eventually the trajectory will have visited all the microstates of the system many times. (*Ergodic* simply means that the trajectory has visited all possible microstates.) Then the frequency of visitation to each box can be assessed.

The other point of the ergodic hypothesis is that we can replace the microstate occupancy frequencies along the trajectory by the relative frequency of occupation of the volumes of the microstates themselves, ignoring the specific trajectory. From this we can construct combinations of microstates and their probabilities of occupancy corresponding to macrostates, where a macrostate is all the microstates to which it corresponds, for example, all the ink particles in a near uniform distribution on a Petri plate. The famous second law of thermodynamics says that the system will flow from macrostates with only a few microstates, such as all the gas particles bunched up in a corner, to a macrostate in which the gas is nearly uniformly distributed. This flow arises because there are many more of the latter kind of microstates in the uniform distribution than in the case of the gas particles bunched into a corner. The increase in entropy is simply a matter of probability: a flow from less-likely macrostates to more-likely ones. This arrow of time, according to many physicists, exists even though Newton's laws are time-reversible. (Recall that I noted concerns about this in chapter 2. Some physicists and philosophers have pointed out that if the directions and velocities of all the particles were reversed from a highly improbable macrostate, the time-reversibility of

Newton's laws implies that the system will again flow to more probable macrostates, so entropy would increase with time running backwards. Subtle arguments about the initial conditions of the universe are sometimes invoked, e.g., by David Albert, to "save" the second law.)

So much for ergodicity. And this is all I shall say about the second law.

Now consider all possible proteins with a length of two hundred amino acids. Typical proteins inside you are about three hundred amino acids long, some several thousand amino acids long, so two hundred is conservative.

If there are twenty types of amino acids in biological proteins, how many proteins like this can there be? This is simple. At each position there are twenty choices, so the total number of possible proteins with a length of two hundred is twenty raised to the two hundredth power, or about 10^{260}. Could all of these proteins arise at least once in the history of our universe since the big bang? The answer is an overwhelming no. Only a miniscule fraction of these proteins can have been constructed in the history of this universe. Thus, as we will see, the proteins that come into existence in the evolution of the biosphere are typically unique—to use Luigi Luisi's phrase, they are never-before-born proteins.

Consider this simple calculation. There are about 10^{80} particles in the known universe. Of course these are widely separated from one another, but let's ignore that inconvenient fact. Suppose all these particles were busy colliding and building nothing but unique, never-before-born proteins of length two hundred ever since the big bang. The universe is 10^{17} seconds old. Suppose it took 10^{-15} seconds, a femtosecond, a very fast chemical time scale, to make each protein.

It turns out that it would require 10^{67} repetitions of the entire history of the universe to create all possible proteins of length two hundred *just once*.

The fastest time scale in the universe is the Planck time scale, 10^{-43} seconds. If proteins of length two hundred were being built on the Planck time scale, it would not change the essential results: now it would take 10^{39} repetitions of the entire history of the universe to create all possible proteins length two hundred once.

This is quite profound. There is no way this universe could have created all possible proteins of length two hundred. The set that actually exists is a tiny subset of the possible. When a new protein is created, say via a mutation, it really is what Luisi happily calls a never-before-born protein. That is, as the biosphere advances into its chemical adjacent possible, it is persistently making unique molecules. Thus as mutations occur so novel proteins are made, or new organic molecules are synthesized in evolution, the biosphere is persistently advancing into its adjacent possible. The adjacent possible is real. We are invading it much of the time.

The same uniqueness occurs at the level of the evolution of species, the human economy, of human history, and human culture. To put the matter simply: we will never explore all the possibilities. *History enters* when the space of the possible is vastly larger than the space of the actual. At these levels of complexity, the evolution of the universe is vastly nonrepeating, hence vastly nonergodic.

It is interesting that these features may apply not only to the biosphere but to other features of the abiotic universe. For example, in chapter 7, I mentioned the giant, cold molecular clouds, the birthplaces of stars, that drift through many galaxies. Complex chemistry occurs in these clouds. For all we know, they are expanding forever into their chemically adjacent possible. Grains that form in the clouds can serve as surface catalysts for further chemistry, new molecules may bind to the grains' catalytic surfaces and modify the subsequent chemical reactions, hence the subsequent flow into the chemical adjacent possible. These grains aggregate up to the size of planetesimals. Almost certainly, they are unique on a grain level, so the giant clouds are almost certainly nonergodic.

Thus at any level of complexity above atoms, from complex molecules upward, the universe cannot make all possible things in many repetitions of its own lifetime. While we are used to thinking of the second law as "the" arrow of time, this, too—the fact that the universe is nonergodic—is an arrow of time with respect to the time-reversibility of the fundamental laws of physics. I explore this issue further just below concerning "underpopulated vast chemical reaction graphs."

Even at the level of simple chemical reactions, the universe appears to exhibit this nonergodicity. Hence the onset of history: what is the behavior of a modest number of atoms on a vast chemical-reaction graph, that is, a small number of atoms on a chemical reaction graph with a hyperastronomical number of chemical species? No one knows for sure, but the answers are likely to change our thinking, as I now discuss.

Consider the familiar concept of chemical equilibrium, when the rate of chemical X converting to chemical Y equals the rate of Y converting to X. At this equilibrium, minor fluctuations occur and dissipate. Suppose that there are very few species involved, X and Y. Suppose further that the total mass in the system, the number of X and Y molecules, is large. That is, there are many copies of each of the two types of chemicals. This is the situation in which normal chemical reaction kinetics and the concepts of chemical thermodynamics were derived. It is a fundamental law of chemical thermodynamics that any such system will approach equilibrium.

Now let us consider a chemical-reaction graph, like that in figure 5.1, but with trillions of billions or more potential organic molecules. For example, we could consider all atoms and molecules made of carbon, nitrogen, sulphur, hydrogen, oxygen, and phosphorus—the normal atoms of organic chemistry. Go to a fine quantum chemist and ask for a list of all possible organic molecules and all possible reactions under defined conditions, up to, say, ten or twenty thousand atoms per molecule. This corresponds to an extremely large reaction graph. We could make the reaction graph indefinitely larger by increasing the maximum number of atoms per molecule to thirty or forty thousand atoms per molecule. By stating the reaction conditions, we are stating which of hugely many reactions can actually occur.

Now the question is this: imagine starting the system with a modest number, say a few thousands or millions, of atoms or molecules at some point or points in this reaction graph and studying the "flow" of the material on the graph. (I should mention that, as far as I know, this exercise hasn't been done.) In many times the lifetime of the universe, it is impossible to make all the possible kinds of molecules with ten or twenty thousand atoms. Further, we are considering the case of a modest number of atoms, far too few to simultaneously form at least one of each of the molecules in the reaction graph. This is radically unlike the case of two molecular species, X and Y, with billions of copies of each kind of molecule. What will happen on our vast, but *underpopulated,* reaction graph?

One possibility is that nothing might happen. Imagine a reaction graph with only two-substrate-two product-reactions—in other words, you need two different molecules to start a reaction. Initiate the graph with mass arranged so that no pairs of substrates are both present for any reaction. That is, for all permissible two-substrate-two-product reactions, only one of the two substrates is present. Without the second substrate, none of the two-substrate-two-product reactions will occur. Notice immediately that, since no reactions occur, the concept of chemical equilibrium as a situation in which the forward and reverse reactions happen equally often simply does not apply. Just by making the reaction graph *underpopulated*, we have called into question, even for a *closed thermodynamic system*, the fundamental laws of such systems, where equilibrium will always be approached. The flow of matter on a vast reaction graph is going to be a complex problem.

What I set out above, with only two-substrate-two-product reactions, is not really a plausible condition. Let's assume that a sufficient diversity of reaction types exist that matter flows from many initial distributions rather easily along allowed reaction pathways. Now start the system with a modest number of atoms or small molecules, define the reaction conditions, and watch the spread of the molecules flowing over the reaction graph (without assuming catalysis). Consider a two-substrate-two-product reaction where there happens to be a single molecule of each of the two substrates and no copies of the product molecules. This specific reaction is perfectly well defined chemically, exhibits what chemists call an enthalpy and an entropy of reaction, and is "shifted to the left"—that is, the substrates are present but the products are not. Thus there is a real chemical energy driving towards the creation of the products into this adjacent possible. In due course such an event may happen.

Now two major points emerge. First, what does the concept of *chemical equilibrium*, even in a closed thermodynamic system with no entry or exit of matter or energy, mean here? It appears meaningless, for the kinds of molecules that arise next can alter indefinitely billions of times over the history of the universe, or billions of lifetimes of the universe. Within the lifetime of our universe, the system never comes to equilibrium. Since this scientific work has never been done, the intuition that chemical equilibrium is here without meaning requires verification.

Second, unlike the square root N fluctuations, which damp out when we consider a system with only a few molecular species, say X and Y, and very many copies of each kind of molecule, here where the number of copies of each kind of molecule is mostly 1 or 0, the *fluctuations presumably do not damp out.* Each time a novel specific set of new never-before-born molecules appear, they create a new adjacent possible into which these new molecules may in turn flow. Fluctuations almost certainly do not damp out. Instead, the specific fluctuations that drive the matter on the reaction graph in particular directions thereby open up new adjacent possible pathways to be explored, possibly creating ever new "salients" of actual molecules on the vast reaction graph. Certainly, the total equilibrium distribution of atoms on all possible molecular species in the graph never remotely recurs. History enters, even in this simple chemical reaction system. Compare this to the square root N fluctuations of the X,Y chemical system that damp out.

Almost certainly, the way the system keeps entering its ever adjacent possible on the underpopulated vast reaction graph creates a nonuniform (more technically, non-isotropic) flow with extra flux in certain directions, which in turn tends statistically to keep wandering off in some historically and contingently biased and unique directions. Fluctuations almost certainly do not damp out. While this kind of chemical reaction system requires detailed study, when the total mass on the graph is tiny compared to the number of molecular species, these guesses seem plausible. In the vast complexity of the giant, cold molecular clouds in the galaxies, with their molecules and grains and planetesimals, something like this may be happening. If so, molecular diversity is exploding in our universe in ever unique ways.

If the system is governed by chemical master equations showing everything that can happen next, and such equations are "solved" by updating the chemical events according to the probability of those events such that each reaction occurs at each moment with some probability, not deterministically, *do we think that we can predict the detailed probabilistic, or stochastic, flow of the matter on the chemical-reaction graph*—even if we could somehow ignore quantum mechanics? I strongly doubt it. Even for small reaction graphs, predicting the *probability distribution* of each of the different possible outcomes requires simulating the system thousands of times from the same initial condition. For an underpopulated but sufficiently vast reaction graph, the simulations could not be carried out by any computer made in this universe. We can call the problem "transcomputational." By making the maximum number of atoms per molecule large enough, far beyond the two hundred amino acids in all possible proteins length two hundred, the number of possible chemical species can be vastly larger than the total number of particles in the universe times the age of the universe on the shortest time scale, the Planck time scale. We cannot know the answers. History and contingency have entered the universe, and we cannot predict what will happen. If we cannot predict even the probabilities of the details of what specific sets of molecules will form in this flow, our incapacity to know what will happen arises well before life arises.

Finally, imagine that some of the molecules can catalyze some of the reactions in the reaction graph. Can the system "stumble upon" an autocatalytic set? If so, what will happen? At least transiently, one might expect that matter will tend to aggregate in and around that autocatalytic set and achieve high concentrations, becoming a macroscopic state of the chemical system.

In summary, we have here the hint of a new arena of standard chemistry on chemical-reaction graphs that are underpopulated by mass. The ideas point to ceaseless novelty.

The flow into the adjacent possible arises at levels of complexity above atoms, certainly for molecules, species, technologies, and human history. Here we must attend to the way the adjacent possible is entered. Salients are almost certainly created in specific "directions" in the space of possibilities, which in turn govern where the system can flow next into its new adjacent possible. These fluctuations almost certainly do not die out, but probably propagate in biased ways into the ever new adjacent possible. All this, as we shall see, applies to the human economy itself as one concrete case. As I shall discuss in detail in chapter 11, there is an economic web of goods and services, where each good has neighbors that are its complements and substitutes. The structure of the web creates ever new economic niches in its adjacent possible for never-before-born, new goods and services, such that economic diversity has expanded. The structure of the economic web partially governs its own growth and transformation.

Just possibly, as hinted in the last chapter, the biosphere and perhaps also the global economy enter the adjacent possible in such a way that the diversity of organized processes in the biosphere, or ways of making a living in both the biosphere and economy, is maximized. To state the matter both cautiously and boldly: Once we consider the nonergodic universe at these levels of complexity, and self-consistently self-constructing wholes like biospheres and a global economy, we are entitled to wonder whether there may be general laws that describe the overall features of such systems even if the details cannot be predicted.

We're only beginning to see the implications of the nonergodicity of the universe. It is part of the persistent creativity in the universe, biosphere, economy, and history. Ceaseless creativity has become physically possible. I believe it is part of reinventing the sacred.

10

BREAKING THE
GALILEAN SPELL

It's time to assess how far we have come. We began this book by looking at reductionism, which has dominated our scientific worldview since the times of Descartes, Galileo, Newton, and Laplace. This philosophy holds that all of reality can ultimately be understood in terms of particles in motion (or whatever the physicists currently suppose is down there), and nothing else has reality, in the sense that it has the power to cause future events. As Steven Weinberg put it, all the explanatory arrows point downward. The logical conclusion is that we live in a meaningless universe of facts and happenings.

What about all the aspects of the universe we hold sacred—agency, meaning, values, purpose, all life, and the planet? We are neither ready to give these up nor willing to consider them mere human illusions. One response is that if the natural world has no room for these things, and yet we are unshakably convinced of their reality, then they must be outside of nature—supernatural, infused into the universe by God. The schism between religion and science is, therefore, in part, a disagreement over the existence of meaning. If meaning were to be discovered scientifically, the schism might be healed.

We have seen that some physicists have come to doubt the adequacy of reductionism, even in physics. We have seen that biological evolution, the coming into existence in the nonergodic universe of organisms and their organization of structures and processes, cannot be reduced, either epistemologically or ontologically, to physics. We appear to be living in an emergent universe in which life and agency arose with no need for a Creator God. With agency, a part of life, comes meaning, doing, and value. Already we are far beyond pure reductionism, into an emergent universe.

But the last chapter hints at something deeper. The vast nonrepeatability, or nonergodicity, of the universe at all levels of complexity above atoms—molecules, species, human history—leaves room for a creativity in the way the universe unfolds at these levels, a creativity that we cannot predict. Even at the level of extremely large chemical-reaction graphs underpopulated with mass, the flow of the chemical system appears to be unpredictable, and fluctuations do not damp out as the reacting system enters the chemical adjacent possible.

In this chapter we will see that Darwinian preadaptations confront us with a radically new kind of unpredictable creativity in the evolution of the biosphere. We will find that we cannot even *finitely prestate* such preadaptations, let alone predict the probability of their occurrences. Thus a radical, and, I will say, partially lawless creativity enters the universe. The radical implication is that we live in an emergent universe in which ceaseless unforeseeable creativity arises and surrounds us. And since we can neither prestate, let alone predict, all that will happen, reason alone is an insufficient guide to living our lives forward. This emergent universe, the ceaseless creativity in this universe, is the bedrock of the sacred that I believe we must reinvent.

Now I want to make my outrageous claim: the evolution of the biosphere is radically nonpredictable and ceaselessly creative. If a scientific law is (as the physicist Murray Gell-Mann concisely put it) a compact statement, available beforehand, of the regularities of a process, then the evolution of the biosphere is partially beyond scientific law. Later we will see that other complex systems share this property. We cannot say beforehand what the regularities of the process are. I cannot over-state how radical what I am claiming is. Since Galileo rolled balls down incline planes and showed that the distance traveled varied with the square of the time elapsed, we scientists have believed that the universe and all in it are governed by natural laws, Newton's, Einstein's, Schrödinger's. Let me call this belief the Galilean spell. Under this spell, we have believed reductionism for over 350 years. I am about to try to exorcize this wonderful Galilean spell that has underwritten so much outstanding science.

Darwinian Preadaptations

We have already discussed Darwinian adaptations as they relate to the heart. As I noted, the heart has several causal properties, including pump-ing blood and making heart sounds. In accounting for the pumping of blood as the "function" of the heart, we imagined Darwin telling us that it was by virtue of this causal property that hearts came into existence in the nonergodic universe. Now, armed with a few additional concepts, we can see that the evolution of the heart was a foray into the adjacent possible. More, the pumping of blood has causal consequences at the level of organisms and the future evolution of the biosphere. Hearts create salients in the adjacent possible that thereafter bias the way evolution occurs. Moreover, while the mechanisms of the heart betray no laws of physics, physics alone cannot predict the coming into existence in the universe of hearts with their particular structure and organization of processes, and with their specific causal powers. Thus hearts are epistemologically and ontologically emergent with respect to physics. Without vitiating any law of physics, physical laws alone do not describe the causal unfolding of the universe. This is already a radical claim, but nothing compared to chal-lenging the Galilean spell.

One of Darwin's brilliant ideas is what is now called Darwinian preadaptation. Darwin noted that an organ, say the heart, could have causal features that were not the *function* of the organ and had no selective signifi-cance in its normal environment. But in a different environment, one of those causal features might come to have selective significance. By "preadapted" Darwin did not mean that some intelligence crafted the preadaptation. He simply meant that an incidental feature with no selective significance in one environment might turn out to have selective significance in another environment.

Preadaptations are abundant in biological evolution. When one occurs, typically, a *novel functionality comes into existence in the biosphere*—and thus the universe. The classic example concerns swim bladders in fish. These bladders, partially filled with air, partially with water, allow the fish to adjust their buoyancy in the water column. Paleontologists have traced the evolution of swim bladders from early fish with lungs. Some of these lived in oxygen-poor water. The lungs grow as outpouchings from the gut. The fish swallowed the oxygen-poor water, some of which entered the lungs, where air bubbles were absorbed, making it easier for the fish to survive. But now water and air were both in a single lung, and the lung was preadapted to evolve into a new function—a swim bladder that adjusted neutral buoyancy in the water column.

With the evolution of the swim bladder a new function has entered the biosphere and universe—that of maintaining neutral buoyancy in a water column. This new function had causal consequences for the further evolution of the biosphere, the evolution of new species of fish and new, never-before-born proteins, and the coevolution of other species with these new species. So the preadaptation changed the future evolution of the biosphere and physical content of the universe in the nonergodic universe above the level of atoms.

Now I come to my radical question. Do you think you could say ahead of time, or finitely prestate, all possible Darwinian preadaptations of all species alive today? Or could you prestate all possible human preadaptations?

I have found no one who believes the answer is yes. We all appear to believe the answer is no.

Part of the problem with attempting to prestate, or specify ahead of time, all possible preadaptations is that to do so we would have to prestate all possible selective environments. Yet we have not the faintest idea of what all possible selective environments might be. More formally, we have no way to list all possible selective environments with respect to all causal features of organisms. How would we even get started on creating such a list? Thus we cannot know beforehand the Darwinian preadaptations that will come to exist in the biosphere. After the fact, once the preadaptation has developed some new functionality, we may well be able to identify it. For swim bladders, the new functionality was the capacity to achieve neutral buoyancy in a water column. Only retrospectively can we sometimes account for the emergence of preadaptations. Paleontologists do this all the time.

But our lack of prior knowledge does nothing to slow down the biosphere's evolution. The profound implication of this is that virtually any feature or interconnected sets of features of an organism might, in the right selective environment, turn out to be a preadaptation and give rise to a novel functionality. Thus the evolution of the biosphere is radically often unprestatable and unpredictable in its invasion of the adjacent possible on the unique trajectory that is its own biological evolution.

This has very important implications. Consider how Newton taught us to do science: prestate the configuration space (say the billiard table); identify the variables, forces among them, and initial and boundary conditions; and solve the dynamical equations for the forward evolution of the system. But we cannot follow Newton's mandate in the evolution of the biosphere, for the very deep reason that we do not know all the relevant variables beforehand. While we know the billiard balls on the table and the table with its boundaries, we do not know beforehand about swim bladders. Therefore we cannot write down the equations among the variables and solve for the forward evolution of the biosphere the way we can for the balls on a billiard table. We are precluded from following Newton. The existence of Darwinian preadaptations means that Newton's way of doing science stops when it comes to the forward evolution of the biosphere.

This, in turn, has implications for whether and to what extent the evolution of the biosphere is describable by "natural laws." The status of laws in science has been much debated among scientists and philosophers of science. Are they in some metaphysical sense "real," in the sense of prescriptively "governing" how things "must" unfold, or are they descriptions? With Nobel laureate physicist Murray Gell-Mann, I adhere to the view that a law is a short, or compressed, description, available beforehand, of the regularities of the phenomenon it covers. But if we cannot prestate, let alone predict, Darwinian preadaptations before they occur, and yet they occur, then we can have no law, in Gell-Mann's sense of a law, for the evolution of the biosphere by Darwinian preadaptations. The same holds true, as we shall see, for the evolution of the economy or for the evolution of human culture. Then Darwinian preadaptations are literally partially lawless, though not in the sense that laws of physics are violated. An oxygen does not change to a nitrogen, but we cannot have a law-governed emergence of swim bladders. A more accurate way of saying this is that Darwinian preadaptations are not sufficiently covered by any natural law. Laws do still apply and are constraints. The oxygen really cannot change to nitrogen. But no natural law governs the emergence of swim bladders in the evolution of the biosphere.

I should stress that in saying that Darwinian preadaptations are not, apparently, law governed, I am not saying that the events that lead to the selection of a preadaptation in a novel environment may not be perfectly causal. That is, without invoking the probabilistic character of quantum mechanics, which may or may not play a role in the emergence of some preadaptations, there may well be perfectly good "classical physical" and biological reasons why and how swim bladders evolved from lungs. Indeed, multiple routes to the evolution of the swim bladder might all have been successful, even if evolution followed only one of these routes. This is another case of the "multiple platform" arguments. Not only can we not finitely prestate all possible Darwinian preadaptations, also for any one, we cannot prestate all possible pathways to its successful emergence in the biosphere. To say that Darwinian preadaptations may be classically caused but are not law governed is quite strange. Indeed, it is radical. Let us call such situations causally anomalous, using a phrase introduced by a philosopher of mind, Donald Davidson, to describe the relations of neural events to mental events. Causally anomalous phenomena are a huge departure from Laplace's reductionism, in which the entire future and past of the universe could be known from the current specification of the positions and velocities of all the particles in the universe. Clearly, we are moving far beyond reductionism.

If the evolution of the biosphere is partially lawless, as I believe is the evolution of the economy and human history, *we are beyond Galileo's spell.* It is not true, it begins to appear, that the unfoldings of the universe, biosphere, and human history are all fully describable by natural law. As we will see, this radical claim has, among its consequences, a radical and liberating creativity in the unfolding of the universe, biosphere, and human civilizations. In this ceaseless creativity in the universe, biosphere, and human culture and history, we can reinvent the sacred, and find a new view of God as the fully natural, awesome, creativity that surrounds us.

I come now to a difficult issue: can I *prove* that Darwinian preadaptations are partially lawless? I do not yet know how to do so. Moreover, it is not at all clear what could constitute such a proof. Let us look for a moment at Gödel's incompleteness theorem. Gödel showed that for any sufficiently rich mathematical system, such as arithmetic, there were statements that were true given the axioms of arithmetic, but that could not be deduced from those axioms. This theorem is stunning, of course. It shows, even for strictly mathematical systems, that they are persistently open in the precise sense that there are true, but not provable statements given the axioms of the systems. And if the true, but unprovable statements are added as new axioms, the enriched axiom set again has new true statements can cannot be deduced from the enriched axiom set. Mathematics itself is not "complete," it is persistently open, as it, by adding new axioms, invades a mathematical adjacent possible.

In a loose sense, Darwinian preadaptations, adding novel functionalities to the biosphere and allowing it, thereby, to invade the physical adjacent possible in new ways in its further evolution, are analogous to mathematics invading the mathematical adjacent possible by adding unprovable statements as new axioms.

In two later chapters, on the economy and mind, I will argue that the analogues of Darwinian preadaptations arise and are not algorithmic. It is not at all clear how to prove that something *is not* algorithmic.

My claim that we cannot finitely prestate Darwinian preadaptations does not pertain to a mathematical world of symbols and equations. It pertains to the real world of atoms and organisms. What could constitute a proof that such Darwinian preadaptations are partially lawless? You might argue that some such natural law might, somehow, be found. We are entering new scientific and philosophic ground, as far as I can tell. I do not know what would constitute a proof of my claim. Since my claim pertains to the real world, not the mathematical world, it appears to be a statement about the real, empirical world, not what philosophers since Hume have called "analytic" statements, such as mathematics, which are not about the empirical world. Hume told us about "synthetic" statements about the real world, such as empirical inductions from specific cases to general laws, for example, Galileo's step from his specific incline plane to a universal law relating distance traveled to the square of the time elapsed. Hume also famously raised the "problem of induction," namely what would justify induction itself? Only induction's past success. But, argued Hume, this is circular, as I discussed above. How then, if at all, could one prove that I am correct, that Darwinian preadaptations cannot be finitely prestated, hence are partially lawless? I can think of no way to do so. Then my way forward is to proceed as I have above, to try to show us that there is no effective procedure to list all causal consequences of parts of organisms and their potential uses, alone or together with other parts of the organism, in all possible selective environments that might arise. This may be all the proof that we can ever have that the evolution of the biosphere is partially lawless. My claim may, itself, be beyond proof.

Whether Darwinian preadaptations are partially lawless, beyond the Galilean spell, they surely occur. Note then, that with respect to Darwinian preadaptations, Weinberg's explanatory arrows do *not point downward to particle physics.* Rather, the *explanatory arrows point upward to the novel selective environment and the preadapted trait* that was thereafter selected to yield a novel function in the biosphere. This is a form of "downward" causation. The selective environment and the causal features of the organism plus natural selection "cause" the emergence in the real biosphere of the new function, for example, the swim bladder maintaining neutral buoyancy in a water column. Downward causation is not mystical, no new forces in physics are called for. Rather, we face here a consequence of the nonreducibility of biology and its evolution to physics alone, and the fact that primitive swim bladders had real causal consequences on the further evolution of the biosphere to yield well adapted swim bladders, that in turn had further causal consequences for the evolution of the biosphere.

Indeed, Darwin's theory and all our knowledge of evolution are at best weakly predictive. We can sometimes predict, under very controlled conditions, what normal adaptations might occur. For example, if you breed a population of fruit flies to increase the number of abdominal bristles, you typically succeed. Darwin's famous finches on the Galapagos Islands are still undergoing normal Darwinian adaptations to reshape their beaks for the already clear function of breaking seeds. But this level of predictability is a vast distance from predicting the evolution of the biosphere, by either normal adaptation or Darwinian preadaptation.

A classical physicist might object. After all, if we take the entire solar system as a huge, $6N$-dimensional classical phase space, the biosphere is just a tiny subvolume of that space. So in a sense we *do* know the classical phase space of the biosphere. In principle, then, the reductionist physicist would say, we *can* list all possible arrangements of particles and their motions that constitute all possible causal consequences of all possible parts of organisms and their environment. Let us grant the physicist's claim. But it does not help at all. *There is no way to pick out in that $6N$-dimensional phase space the "collective variables" or "degrees of freedom," such as swim bladders, lungs, or wings, that are preadaptations and will have causal roles in the future evolution of this specific biosphere.* Once again, we seem firmly precluded from writing down the variables and laws, and solving the equations for the forward evolution of the biosphere. We cannot follow Newton. No natural law governs the emergence of Darwinian preadaptations. So we cannot write down that law and solve for its dynamical consequences in the evolution of the biosphere.

Even if we lack equations for the collective variables such as lungs and their causal roles, could we simulate the evolution of this specific biosphere? As we know, organisms straddle the quantum-classical boundary. Cosmic rays can cause mutations that alter proteins and confer evolutionary advantages or disadvantages. If we also assume, as do almost all physicists, that spacetime is continuous, then it would seem to take an infinite number of simulations, or even a second-order infinite number of simulations, to cover all possible quantum events in an infinite set of simulated biospheres. It's simply not possible to do this simulation. A critical reductionist physicist might object that we don't really have to carry out all the second-order infinity of simulations. Small enough differences in the circumstances of quantum events may not matter. We can "coarse grain" the spacetime continuum and classical and quantum processes within it. Hence there are a vast, but finite, number of simulations of possible biospheres. But even here, the physicist cannot know beforehand which of the second-order infinite number of simulations can be *safely skipped and still simulate the evolution of this specific biosphere in one of those simulations.* That is, any coarse graining of space-time and events has grain boundaries, and we cannot know beforehand where it is safe to place those boundaries on continuous spacetime and the events. Further, even if we could carry out the vastly many but finite coarse grained simulations, there is simply no way to verify for any specific simulation that all the throws of the quantum dice were the exact ones that actually occurred in the 3.8 billion years of evolution. For all we know, slight differences in the quantum events might have caused the biosphere to veer off in radically different directions. Different mutations might have occurred. Thus, even if we could do the simulations, we could never know which of many such simulations captured the specific biosphere that we inhabit. This point remains valid even if spacetime is not continuous, say on the smallest space scale, the Planck scale of 10 to the −33cm. Now the number of simulations is gargantuan, but finite without arbitrary coarse graining. Yet again we cannot know, or confirm or disconfirm, which history of "a" biosphere that we simulate maps to the specific evolution of *this* specific biosphere.

In critical summary, we truly cannot know ahead of time the way the biosphere will evolve. Then, if this radical new view is correct, we have stepped beyond the Galilean spell. The evolution of the biosphere really is partially lawless. More, Darwinian preadaptations appear to preclude even sensible probability statements. Consider flipping a fair coin ten thousand times. It will come up heads about five thousand times, with a binomial distribution about that mean. Here we confront the familiar "frequency" interpretation of probability. The expected frequency of heads is 0.5. But notice that a statistician knows in advance all of the possible outcomes of flipping the coin a thousand times. She knows the full configuration space of possibilities, and bases her prediction of likely outcomes on that knowledge. In statistical language, she knew the "sample space" beforehand. But had we been able to observe the evolution of the lung to the swim bladder, for example, we would have had no idea beforehand that it might happen. We do not know the full space of possible novel functionalities that might arise, and therefore we cannot form a frequency based probability statement about Darwinian preadaptations. Again, in statistical language, we have no idea of the sample space, so cannot make probability statements.

There is a second interpretation of probability, initially due to Laplace, that is the basis of Bayesian probabilities. Suppose, said Laplace, that I am before N doors, behind one of which is a fortune. I have no idea which door hides the treasure, and I have to guess. What is the probability I will succeed? $1/N$. Here there is no frequency interpretation, but notice that I know ahead of time the full set of doors, N. I do not know this with respect to Darwinian preadaptations of possible novel functions. I do not know N. I do not know the sample space ahead of time. So neither interpretation of probability can be applied to Darwinian preadaptations. Hence it seems we cannot predict such preadaptations, nor even make sensible probability statements about them.

Now that we understand the rudiments of Darwinian preadaptations, I can make the breaking of the Galilean spell still more radical. It is a known fact that the three middle ear bones that allow you to hear evolved by Darwinian preadaptations from the jaw bones of an early fish. Thus, multiple parts of the fish, the three jaw bones, came together to jointly create a new function relevant to hearing. In chapter 12, Mind, I will discuss, with some amusement, whether we can prestate all the possible uses to which screw drivers might be put in all possible circumstances. I hope you will be convinced that we cannot do so. Now imagine several "parts," a screwdriver, a nail, and cloth. Can we hope to list all possible ways these might be used together for all possible functional uses in all possible circumstances? I think the answer is clear that we cannot, for we cannot list all the uses of any of the components, let alone their uses, jointly. But the screwdriver, nail, and cloth are analogous to the three bones in the fish jaw that evolved by preadaptation into the bones of the middle ear. Even more radically, the properties of the screwdriver, nail, and cloth might include all possible *relational properties,* such as the distance of the screwdriver from the nail and a nearby wall. So, too, with the evolution of Darwinian preadaptations; they include relational properties, such as the proximity of the three fish jawbones to one another in their evolution to the three bones of the middle ear. Yet any of these properties, nonrelational and relational, could become the basis of a Darwinian preadaptation. I conclude that we cannot prestate all these possibilities, hence that Darwinian preadaptation, and their analogues in the economy discussed in the next chapter, are partially lawless. We are beyond the Galilean spell.

I need to point out that what I am saying is radically different from what is called deterministic chaos, where a very important kind of unpredictability occurs. These systems are well understood. If one writes down certain differential equations for systems with three or more variables, such as models of the weather, then the "attractors" of such dynamical systems can be "chaotic." Here what is meant is that the system converges in its state space onto one or more chaotic attractors, but once on the attractor, the most minute changes in the state of the system induce amplifying divergence in the future dynamical behavior of the system. This is the famous "butterfly effect," where a butterfly in Paris causes the weather in Chicago to change. More formally, the butterfly effect is called "sensitivity to initial conditions." Now it is entirely true that the future evolution of a chaotic system cannot be predicted without infinite precision in knowledge of the initial conditions, and such infinite precision cannot be obtained. So prediction of the long term behavior of the system on its chaotic attractor in detail is truly precluded, given finite precision in knowledge of the initial condition.

I now need to show you that this incapacity to predict in deterministic chaotic systems is emphatically *not* the same as the failure to prestate or predict Darwinian preadaptations. In the deterministic chaotic case, *we know beforehand the state space of the system,* in the simplest case, three continuous variables and their ranges. But in sharp contrast, we do not know beforehand the state space, or sample space, of the evolving biosphere and the emergence in the nonergodic universe of swim bladders. Or, if we attempt the reductionist stance, while we might know the state space in the sense of the classical $6N$-dimensional phase space of the solar system, or its analogous quantum state space, called Hilbert space, we have, as we have seen, no way to pick out the relevant collective variables that will play a causal role in the further evolution of the biosphere, or to confirm or disconfirm which of vastly many or infinitely many simulations of biospheres correspond to the evolution of this specific biosphere.

Deterministic chaos does not break the Galilean spell. It is fully lawful. Darwinian preadaptations do break the Galilean spell. We have entered an entirely new philosophic and scientific worldview after almost four hundred years. This is, indeed, radical. From it spreads a vast new freedom, partially lawless, in the evolution of the biosphere, economy, and human history in this nonergodic universe.

Clearly, we are a very long way from Weinbergian reduction. Not only is the Galilean spell broken, but the explanatory arrows with respect to the emergence of novel functions point upward to the evolutionary emergence of preadapations via natural selection for nonprestatable functions in nonprestatable selective environments. They do not point downward to string theory. We have seen that the evolution of life violates no law of physics but cannot be reduced to physics. We are well into a new scientific worldview in which we are members of a universe of creativity and history, a stunningly creative biosphere which we cocreate. In this new scientific worldview, our place in the world is very different from that envisioned by pure reductionism. These, I hope, are long steps towards reinventing the sacred. If we contemplate the diversity and complexity of the biosphere, and the human mind, consciousness, economy, history, and culture—our emergent historicity—the ceaseless, partially lawless, ever creative exploration of the adjacent possible on all these levels, how can we be less than awestruck?

To summarize, we have reached one of the central points in this entire book. Since Galileo discovered a quantitative relationship of a ball rolling down an incline plane for an interval, T, and the distance it went, we have been under the Galilean spell that all is governed by natural law. It is no accident that Galileo got in trouble with the Church, not for his heliocentric, Copernican views, but for believing that science, not revelation, was the only true path to knowledge. From Galileo we come to Newton, to Laplace, to Schrödinger, to reductionism, to the view that, when we understand it all,

"all" will be covered sufficiently by natural law. This is the Galilean spell that holds us in its sway. With it, spirituality seems pointless, as does the universe of the reductionist. But if Darwinian preadaptations cannot be finitely prestated, we move not only beyond reductionism to emergence, as I have discussed so far, we also appear to move, for the first time in four hundred years, beyond the spell that brilliant Galileo cast over us all: that "all" would someday be covered by sufficient natural law. In its place we will find a profound partial lawlessness as we invade the adjacent possible in the nonergodic universe. With it we will find ceaseless creativity in the universe, biosphere and human life. In that creativity we can find one sense of God that we can share. This is, I believe, the core of why we have wanted a supernatural God. Such a God may exist, but we do not need that supernatural God. The creativity in nature is God enough. From this natural sense of God, we can hope to reinvent the sacred as the creativity in nature—without requiring that all that happens is to our liking. From that new sacred, we can hope to invent a global ethic to orient our lives, and our emerging global civilization.

There is, before us, another enormous issue. It bears on the global civilization that will emerge and that we will jointly partially co-construct. If Darwinian preadaptations are partially lawless, what are we to make of the emergence and evolution of the biosphere? No one is in charge. Some of the processes, such as preadaptations, are partially lawless, species coevolve with one another and the abiotic environment they sometimes modify, and cocreate an evolving whole. How can such self-consistent, self-constructing, partially lawless wholes come to exist and evolve in the universe? We will confront much the same issues with respect to the economy, discussed in the next chapter. And the same issue confronts the evolution of human history and civilization, including a forthcoming global civilization. But with the economy, history, and evolution of a global civilization, we become puzzled by the roles of human conscious intention in the evolution that occurs. With the biosphere the issue does not arise, certainly before about a billion years ago, before we think any living thing was conscious. Somehow, despite an absence of sufficient law on the lower level, Darwinian preadaptations in the case of the biosphere, a self-organized, evolving biosphere emerges and has continued literally to co-construct itself for 3.8 billion years. How does the whole remain, in some sense, coherent such that species continue to be able to coexist, even as extinction events happen and new species evolve? There appear to be missing principles of coevolutionary assembly for the biosphere, the human economy and civilization, and even evolving human law. In the next chapter, I will introduce a theory called self-organized criticality, and show that it may provide laws for the evolution of the biosphere and economy, despite absence of law at lower levels. The deep potential implication is that laws may emerge on higher levels where no laws exist on lower levels. This is not so far from R. Laughlin's views in *A Different Universe: The Universe from the Bottom Down*. How can we not be stunned by the fact that the biosphere, economy, and human cultural evolution are co-constructing wholes despite partial lawlessness? We are, indeed, beyond the Galilean spell, into creativity and emergence, to the unknown mystery into which we must live our lives.

AN ALTERNATIVE VIEW: INTELLIGENT DESIGN AND AN UNSPOKEN SUPERNATURAL GOD

There is a new pattern of reasoning about evolution that would deny the ceaseless creativity in the biosphere of which I speak. This pattern of reasoning is called intelligent design, is driven largely by Christian religious fundamentalism, implicitly seeks a need for a Creator God—although it does not admit this—and is not science. I will, however, give it the benefit of the doubt below, and conclude that if intelligent design can actually be construed as putative science, it stands, at present, disconfirmed.

Intelligent design (ID) has received a great deal of attention in the last few years. As most readers know, it takes the view that certain features of life are so irreducibly complex that they cannot have arisen through natural selection and so must have been designed. In this book I am concerned mainly with the scientific status, if any, of intelligent design. Darwinian preadaptations are central to the biologist's response to ID.

First, we need to be accurate about the religious roots of intelligent design. It grew out of a more primitive idea called creation science, whose proponents wished to show that the Abrahamic biblical tradition was factually correct. The religious origin of creation science is beyond dispute, and several efforts to have it taught in public schools alongside standard evolutionary theory failed in the United States courts. The founders of creation science, all deeply opposed to both evolution of life on earth and Darwinian selection in particular, regrouped, added new members, and devised intelligent design. It has been an astonishing political success. Members of the United States Congress espouse ID. In 2006, a U.S. district court in Dover, Pennsylvania, struck down instruction of intelligent design in public schools as an unconstitutional attempt to breach the division between church and state, in part because a relevant ID book was proven legally to have originated as a creation science book.

Still, the religious roots of intelligent design should be viewed with empathy. For those who believe in a Creator God as author of the universe and of moral law, the profound fear behind the attack on evolution is the fear that without a Creator God, the roots of Western civilization will crumble. We will be left with a Godless, meaningless, amoral secular humanism. One deep purpose of this book is to say to those who hold this view and are driven by this fear, that if we take God to be the creativity in the universe and find, as we shall, the roots of our ethics in higher primates and other animals due to evolution itself, then this deep fear is unfounded. God as the creativity in the universe can, I believe, offer us a view in which the sacred and the moral remain utterly valid. So I want to say that I am sympathetic with the feelings and beliefs of those who espouse intelligent design. But as science, it fails. More, we do not need a supernatural Creator God, we need to break the Galilean spell.

What, then, is the status of intelligent design as science?

I should first point out that when politicians and religious leaders support apparent science, and virtually no scientists do, it is a beginning clue that science itself is not leading the investigation. And in fact, the vast majority of scientists simply dismiss intelligent design. If not for its political visibility, they say, it would be hardly worth refuting. I shall take a gentler stance. I believe that intelligent design could be legitimate science. What I will say is that, insofar as intelligent design wishes to be taken as real science, *it does make a prediction and that prediction appears to be false.* Thus if intelligent design purports to be science, it stands largely disconfirmed.

Yet here is why intelligent design could be science. Suppose we received a "message" from space that ET was home. There is a lot of noise and signal in the universe that might reach our detectors. How would we decide that the "message" was not 100 percent noise? This in itself is a legitimate scientific question. In fact, if SETI, the effort to find such signals, ever claims to succeed, you can be sure that their criteria for asserting that the message was intelligently designed and not random will be widely scrutinized. Thus, "designed or not" is a real question.

Here is the cornerstone of the intelligent design movement. The claim is that some biological features are irreducibly complex in a way that requires appeal to design. A favorite example is the flagellar motor in a bacterium. This motor, which is indeed remarkable, consists of an electric engine with rotor and stator, and a variety of components that jointly allow the motor to turn the flagellum clockwise or counterclockwise, driving the bacterium up the glucose gradient. There is no way, assert ID's proponents, that natural selection can build such a device. From what precursors could it possibly have evolved? Remove any component and the whole ceases to be an effective motor.

The beginnings of an answer to the ID movement came in a humorous talk by a scientist court witness in Dover who appeared with a mouse trap broken into parts. He used one part as a tie clip. Another part was used as a pad to hold paper, and so on. The visual point was that parts of the assembled mouse trap could be used for other purposes than constituting a mouse trap. The parts might later then be assembled into one machine with the novel function of trapping mice.

You will see immediately one of the standard biologist's responses to irreducible complexity—Darwinian preadaptations. We do not have to assume that the flagellar motor was assembled as a motor from its evolutionary start. Rather, parts selected for a variety of different initial functions might have come to have yet further novel functions, until the flagellar motor was in the adjacent possible and could be assembled by an evolutionary process involving yet a further Darwinian preadaptation. Indeed, the scientists in the Dover court case showed examples of partial flagellar motors in other bacteria used for other functions. The flagellar motor is virtually certainly a Darwinian preadaptation, like the three bones of the human middle ear.

Now in advocating this view, the biologist faces two issues: do Darwinian preadaptations occur frequently enough to be significant? Here the answer is yes, with evidence to support it. Second, in any specific case, is the sequence of preadaptations known that led ultimately to that specific complex feature of an organism? Sometimes the answer is yes, we know the sequence of adaptations and preadaptations in question. Often, the answer here is, "No, at least not yet known." Unlike creation science, intelligent design does not deny the process of evolution. Proponents of ID must therefore admit that complex new functionalities can arise through preadaptation. After all, swim bladders did evolve from lungs, and the human middle ear bones do derive from jaw bones of an early fish. The strong prediction of ID, then, must be that no Darwinian preadaptations occur that might, by a succession of novel functionalities, give rise to an irreducibly complex whole. But this prediction is almost certainly false. Preadaptations occur, and reasonably often at that. Such a string of preadaptations is the obvious evolutionary explanation of complex wholes. Thus if intelligent design is to be taken as a science, and must make this prediction, it is almost certainly disconfirmed.

In addition, intelligent design is based on probability arguments. It says that the flagellar motor, for example, is too improbable to have arisen by chance. It is irreducibly complex and so improbable that there must be a designer. But we saw above that we cannot make probability statements about Darwinian preadaptations, for we do not know beforehand the full configuration space. ID simply cannot compute that a given irreducibly complex entity such as the flagellar motor could not have come about by a sequence of Darwinian preadaptations in reasonable time. Its probability calculations are entirely suspect. The sample space is, again, not known beforehand.

Intelligent design enthusiasts might try to wiggle out of the prediction that there are no Darwinian preadaptations accounting for a given complex whole by saying that in the case of a specific irreducibly complex entity such as the flagellar motor, there happened to be no Darwinian preadaptations that arose on the way to its evolution. This step has problems, but note first that it is a completely falsifiable prediction in the Popperian sense. We need only find one or more preadaptations on the evolutionary path to this specific complex entity. And as noted, such a partial motor used for another purpose has, in fact, been found in another bacterium. So ID's favorite case of the flagellar motor is at least partially disconfirmed.

But I am not a Popperian. The philosopher Karl Popper argued, to overcome Hume's problem of induction, that scientists do not "induce," rather we boldly come forward with hypotheses and then set about trying to prove them false. A large part of what drove Popper was the observation that the universal claim "All swans are white" can never be proven since we can never find all swans. But it can be disproved by a single black swan. Even a beige one will do. Popper reasoned that it was easier to disprove a bold hypothesis than to prove its universality, so to segregate science from nonscience he held that scientific hypotheses must be falsifiable.

It is an amusing fact that scientists who eschew philosophy invariably espouse a philosophy of science that is long outdated. Most scientists today will somberly argue that hypotheses must be falsifiable. But science and real life are more complex. The Harvard philosopher W. V. O. Quine advanced the holism in science thesis. Suppose I hold that the Earth is flat. You know the Earth is an oblate spheroid. We conceive a test: We will go down to the seashore and watch a sailing ship sail away from the shore toward the horizon. We reason that if the Earth is flat, the ship should dwindle to a point. If the Earth is a sphere, the hull should disappear over the horizon before the sails. We sit down with sandwiches and a fine wine-and-cheese combination and watch the ship sail away. Sure enough, it is hull down over the horizon. "Fool," you tell me. "The Earth is a sphere. You pay for lunch." "Not so fast," I reply. "Maybe the ship sank!" Thus, said Quine, given disconfirming evidence, we can always doubt the facts.

So we then radio the ship and learn it is underway. "Now admit the Earth is a sphere," you demand. "Not so fast," I reply. "You have assumed that light rays travel in straight lines in a gravitational field. I think this experiment shows that your assumption about light rays is false. Light rays fall in a gravitational field and those from the bottom of the boat hit the water first, so the hull disappeared first. You pay for lunch."

"Well," says Quine, "I have a point. No hypothesis confronts the world alone. It always arises in a web of other statements of fact—the ship did or did not sink—and a web of other laws or assumptions about the world. Given negative evidence, we have to give up something, but we are not forced to give up any specific hypothesis, such as our hypothesis that the world is flat, as long as we change something else. We might, however, wind up with a very awkward physics."

"So," asserted Quine sensibly, "what we actually do is provisionally alter those statements of fact or other laws that minimally alter our worldview." This is Quine's holism in science. And it is broader than science. In general we have a web of beliefs and alter them piecemeal—except for true revolutions like general relativity—so that we minimize the total disruption to the web of ideas. Any historian of science or culture would add that social factors enter in these choices as well.

This is directly related to the intelligent design movement. Suppose we really concluded that the facts of biology require a designer. This would be a stunning revolution. It would demand overwhelming evidence. And it would require an overwhelming problem to which only ID, not preadaptations, could be the solution. Like general relativity, it would call for a variety of further testable predictions. Therefore, ID is disconfirmed by Darwinian preadaptations of the only prediction that I can think of that it can make—even though I am not a Popperian. Its thesis is so overwhelming, if true, that it will require extraordinary evidence to support it, and Quine's holism in science thesis powerfully suggests that it is much less of a disruption to our entire understanding of the real world to invoke Darwinian preadaptations, not ID, to explain the flagellar rotary motor.

There is a final point to make. If intelligent design were serious science, one would think that the questions of who or what the designer is, and how the designer actually manages to assemble the irreducibly complex wholes, would be legitimate questions. These remain unasked because, of course, the designer is, carefully unspoken, the supernatural Abrahamic God who can act in the world by miracles to create the complex wholes.

Given that intelligent design makes some statements that qualify as scientific, should it be taught as science? Of course not. As science, it simply leads nowhere. And the court properly recognized that intelligent design is religion in disguise.

We are beyond the hegemony of the reductionism of half a century ago. We have seen that Darwinian natural selection and biological functions are not reducible to physics. We have seen that my law of collectively autocatalytic sets in the origin of life is also not reducible to physics. We have seen creditable evidence that science is moving forward towards an explanation for the natural emergence of life, agency, meaning, value, and doing. We have, thus, seen emergence with respect to a pure reductionism. Thanks to the nonergodicity and historicity of the universe above the level of atoms, the evolution of the biosphere by Darwinian preadaptations cannot be foretold, and the familiar Newtonian way of doing science fails. Such preadaptations point to a ceaseless creativity in the evolution of the biosphere. If by a natural law we mean a compact prior description of the regularities of the phenomena in question, the evolution of the biosphere via preadaptations is not describable by law. We will soon find its analogues in economic and cultural evolution, which, like the biosphere, are self-consistently self-constructing but evolving wholes whose constituents are partially lawless.

This is a radically different scientific worldview than we have known. I believe this new scientific worldview breaks the Galilean spell of the sufficiency of natural law. In its place is a freedom we do not yet understand, but ceaseless creativity in the universe, biosphere, and human life are its talismans. I believe this creativity suffices to allow us to reinvent the sacred as the stunning reality we live in. But even more is at stake. Our incapacity to predict Darwinian preadaptations, when their analogues arise in our everyday life, demands of us that we rethink the role of reason itself, for reason cannot be a sufficient guide to live our lives forward, unknowing. We must come to see reason as part of a still mysterious entirety of our lives, when we often radically cannot know what will occur but must act anyway. We do, in fact, live forward into mystery. Thus we, too, are a part of the sacred we must reinvent.

11

THE EVOLUTION
OF THE ECONOMY

Let us turn to human action. In this chapter, I will move beyond the "hard" sciences and show how the ceaseless creativity of the universe is manifested in a particular social realm, the realm of economics. Like the biosphere, our human realm is endlessly creative in ways that typically cannot be foretold. Our common incapacity to know beforehand what Darwinian preadaptation will bring to biological evolution has an analogue in technological evolution—we can easily apply the idea of preadaptation there, too—and more broadly in the evolution of the economy. Later in this chapter, I will discuss economic webs among goods and services, and show how that web creates novel niches for ever newer goods and services. The global economy has exploded from perhaps a hundred to a thousand goods fifty thousand years ago to an estimated 10 billion today. The economic web invades its adjacent possible.

Like the biosphere, the "econosphere" is a self-consistently co-constructing whole, persistently evolving, with small and large extinctions of old ways of making a living, and the persistent small and large avalanches of the emergence of new ways of making a living. Because there are economic analogues of preadaptations, I hope to persuade you that how the economy evolves is often not foreseeable. Despite this, the evolution of the economy often remains whole as it evolves and adapts. On a broader scale, the same self-consistent co-construction and adaptive evolution to that which is unforeseeable is true of our civilizations, including the global civilization that may well be emerging and which we will, as humans, partially co-construct.

This discussion is not only scientifically important but critical in the practical world. The structure of the economic web that I will describe below influences its role in driving economic growth. I will show here that the growth of the web is self-amplifying, or autocatalytic, and that—as data confirm—economic growth is positively correlated with economic diversity. The more diverse the economic web, the easier is the creation of still further novelty. This has practical implications. Despite successes in achieving economic growth in Asia, including China, India, Korea, Taiwan, and Singapore, and in eastern Europe including Russia, much of the world today still lives in poverty. Understanding the economic web and its modes of evolution may perhaps alleviate some of that poverty while achieving ecologically sustainable growth. I will argue here that there are severe limits to this ambition: our incapacity to predict technological evolution and other aspects of the evolving economy is an implicit condition of the creative universe. If we cannot extend science, Newtonian or otherwise, to the detailed evolution of the economic sphere, we must reexamine rationality in a broad framework. We must strive to understand how fully human agents manage to live their economic lives in the face of inescapable ignorance, thus where reason alone is an insufficient guide to our actions. We will have to re-examine how we use all the tools that we have evolved for 3.8 billion years, of cellular evolution, animal evolution, vertebrate evolution, mammalian evolution, and hominid evolution, to see our full human selves living our lives.

Darwinian Preadaptations in Technological Evolution

The following story is said to be true: A group of engineers were trying to invent the tractor. They knew they would need a massive engine. They obtained a massive engine block and mounted it on a chassis, which promptly broke. They tried successively larger chassis, all of which broke. At last one of the engineers said, "You know, the engine block is so big and *rigid,* we can use the rigidity of the engine block itself as the chassis and hang the rest of the tractor off the engine block." And indeed that is how tractors are made.

Now the rigidity of the engine block was an unused causal feature of the engine block that came to be used for a novel function—to serve as a chassis. It is a Darwinian preadaptation in the economic sphere: legally, it was an invention, presumably patentable.

Can we prestate all technological preadaptations? In the extreme, obviously not: no one in Roman times foresaw cruise missiles. Even shortly after the invention of the computer, Thomas Johnson Watson Sr., the president of IBM, thought there would be a global market for perhaps nineteen IBM 701s, no one foresaw the Internet and World Wide Web.

More generally, the economist Brian Arthur points out that most inventions wind up being used for purposes unforeseen at the time of the invention. The use of the computer as a word processor is a familiar example. Such novel uses are precisely parallel to Darwinian preadaptations. With respect to both the frequency interpretation of probability and Laplace's N door interpretation of probability, we appear to be precluded even from making sensible probability statements about the emergence of novel functions in the economy. As I described in the previous chapter, in the frequency interpretation of probabilities, we might toss a fair coin ten thousand times and predict that it would come up heads about five thousand times. But it is essential that we know, beforehand, the "sample space" of all the possible outcomes of tossing the coin ten thousand times. In Laplace's example, we are confronted by N doors behind one of which is a treasure. We have no idea which door hides the treasure. Laplace tells us that the chance we pick the right door is $1/N$. This is not a frequency interpretation of probability, but notice again that we know beforehand the sample space; it is the number of doors, N. In Darwinian preadaptations, biological or economic, we typically have no idea at all what the sample space is, what the possibilities are, before they are evolved or invented. By and large, we cannot make well-founded probability statements about the probability of the Internet in 1920, or 1840.

In short, typically we know little, even about the near future. This is, of course, an overstatement. The venture capital community would not exist if there were no investors with foresight and a capacity to assess the potential market for novel goods and services. But such judgments rely as much on intuition as on strict algorithms, that is, effective procedures to compute the probabilities of unknowable outcomes and inventions and optimize with respect to those probabilities. And even then, the best venture capitalists are more often wrong than right. Out of ten companies which receive venture capital funding, perhaps one will succeed. The others fail for a variety of reasons, including bad management and the fact that the business did not, in reality, have the promise that was supposed.

Below I will discuss the slight possibility that we could simulate the evolution of the economy. To do so, we would have to be able to prestate all possible novel functionalities, such as using the engine block as a chassis. I doubt that this is possible. I believe there is no lowest-level, basement language of a set of simple functionalities from which all higher functionalities that might ever be useful in the evolving economy can be derived logically. As I will discuss below, it does not appear that we can, from the properties of Lego blocks in themselves, even list all possible uses, or functionalities, of Lego block cranes and rolling platforms, let alone real economic goods such as computers, now unexpectedly used as word processors. We cannot simulate the evolution of the economy by preadaptations. In addition, of course, the evolution of an individual company is subject to vicissitudes ranging from access to capital to start a venture, to exogenous shocks, to management failure.

Despite unprestatable economic preadaptations, the economy does often evolve in ways that look predictable. Just as there are normal Darwinian adaptations like the evolution of beaks in Galapagos finches, there are "normal" adaptations in technological evolution. Some of it is merely improvement on a function already in hand. For example, there are the famous "learning curves" in economics. A typical illustration of the learning curve is that in a factory making trucks, every time the number of trucks manufactured doubles, the cost per truck falls by a constant fraction. Companies count on going "up" such learning curves through accumulated alterations in procedures as they make more of their product. Learning curves are ubiquitous. So the claim that we cannot anticipate all technological evolution does not mean that we cannot anticipate learning curves. Thus, we often have considerable near-term insight into the expected improvements in technologies with known functions.

In my book *At Home in the Universe* and in subsequent economic publications with colleagues, I discuss the possibility that the statistics of learning curves are due to "hill climbing" on more or less rugged payoff landscapes. For simple models of such landscapes, which derive from a form of disordered magnetic material called "spin glasses" in physics, it is possible to build models that capture the statistics of learning curves as more or less "myopic" hill climbers search for local or global peaks in the payoff landscape.

There is beginning to be good evidence for both normal and Darwinian preadaptations in economic evolution. My colleague Ricard Sole has examined millions of patent citations and finds evidence both for normal adaptations and abundant Darwinian preadaptations of novel functionalities. We have to develop refined criteria to distinguish these, in kind or in degree, for our further economic analysis of their roles in technological evolution.

In its classic self-definition as "the study of the allocation of scarce resources that have alternative uses," economics sets its sights too low. The economy is not just the allocation of scarce resources. The ambitions of the field of economics should include an adequate theory, history, and understanding of the explosion of goods and services in the past fifty thousand years, and of novel ways of making a living. Both explosions are real. As they occur, they persistently create the conditions for the next new ways of making a living with new goods and services, and passing of old ways of making a living with old goods and services. The economy, with us human agents participating in it, truly is a co-constructing, partially unforeseeable, evolving system. Although there are economists who discuss these issues, most have no theory of this explosion. Yet it is perhaps the stunning factor driving the growth in economic wealth globally in the past fifty thousand years. Compare our lives in the first and even the third world today with that of *Homo sapiens sapiens* of the lower Paleolithic. If we understood this explosion, we might know better how to help the evolution of wealth.

The central concern of modern economics is how markets match supply with demand. Kenneth Arrow and Gerard Debreu won the Nobel Prize for their elegant theory describing this. Imagine, they said, all possible "dated contingent goods." An example of a dated contingent good is a basket of wheat, delivered next month provided I grow a beard. The contingency is my growing a beard. *Given the prestated set of all possible dated contingent goods,* Arrow and Debreu imagine an auctioneer who gathers us all together. Based on our perfect rationality, we bid for contracts on all possible dated contingent goods, basing our bids on our estimates of the future probability of the contingency of each dated contingent good, and its utility or value to each of us. When the auction is done we sign our contracts and go home. In a brilliant mathematical theorem, Arrow and Debreu prove that however the future unfolds, our respective contracts fitting those of the dated contingent goods contracts that actually will come due are executed, and all markets will clear, meaning that all goods will be exchanged, supply will match demand, and some value will be received by everyone. This market clearing is known as equilibrium, hence the theory is called competitive general equilibrium.

This theory is the foundation of economic thinking today. It is beautiful. Notice that it is not reducible to physics, because the economic rational agents have foresight and use their knowledge to calculate the probabilities of each dated contingent good, for example that I will grow a beard and that a bushel of wheat is available for sale next month. In addition, the agents have *utilities*. That is, each dated contingent good has some value to each of the agents. Physics has only events, no values, and surely no foresight. Because values enter the universe with agency, and agency is part of life, and the evolution of life cannot be reduced to physics, as I discussed in chapter 4, economics cannot be reduced to physics.

Despite the beauty of the Arrow-Debreu theory, in the real world of unforeseeable technological evolution we can neither prestate all possible dated contingent goods nor even form a probability assessment of those yet to be invented. If we somehow manage to be "rational" about technological preadaptations, it is not by any calculus of probability. Thus the Arrow and Debreu theory is of limited applicability in the real world. We don't know all possible goods ahead of time.

This tells us that competitive general equilibrium, as an attempt to account for market clearing in the real world, either needs to be limited to short time horizons with respect to technological innovations, or expanded in some unknown way. Since we cannot even simulate the technological evolution of the economy, because we do not know what economic Darwinian preadaptations will arise in its evolution, as discussed further below, competitive general equilibrium can only function as a short-term approximation, perhaps to some new theory. But I strongly suspect we will find that we come here, too, to a limit of standard science. I do not think we can deduce the detailed evolution of the economy.

There are two other major strands to the core of standard contemporary economics: game theory and rational expectations. Both fail to predict real economic behavior, and for similar reasons.

In game theory, a set of players have a set of predefined strategies, and a payoff for each player that depends upon his or her strategy and the strategies of all the other players. But again, the set of strategies and payoffs is predefined. This is simply not true in the real economic world, where unpredictable new goods and services arise at unpredictable intervals. At present it is entirely unclear how to extend game theory to include the possible payoffs for novel goods and services we cannot foresee, and about which we cannot make probability statements. By what calculations would economists "optimize" the behavior of a rational economic agent? As Niels Bohr said, prediction is so difficult, particularly about the future. If we cannot be pure rationalists, then we will have to return to the issue of how we live our lives in the face of the unknown. Somehow we do so, with faith and courage, with all our evolved sensibilities intact. Here in considering how human agents act in the real world of business we confront for the first time in this book the reality of the need to understand our full humanity, split asunder since the days of the metaphysical Elizabethan poets.

The theory of rational expectations makes the same assumption as the other two theories: that there is a prestatable set of goods and services. This elegant theory then asserts that the economy can behave in ways not predicted by competitive general equilibrium if all the economic agents are hyperrational, and if they share a set of beliefs about how the economy will behave that has the property that if they each optimize their behavior under those beliefs, that behavior will self-consistently "instantiate" the economy in which the agents believe. To oversimplify, it is a theory of economic behavior in a world of self-fulfilling prophesies. Rational expectation powerfully and elegantly explains, for instance, why there can be speculative bubbles, which are precluded under competitive general equilibrium.

There are, however, at least two major problems with the theory of rational expectations. Because goods and services evolve in ways that are often the analogues of Darwinian preadaptations and cannot be foreseen, expectations based on a given set of goods and services may be wrong. Second, as I discuss in *Investigations* in a simple model, even if the goods and services are assumed to be fixed, the rational expectations' "equilibrium" is unstable. Strategic economic agents will instead create an unstable set of coevolving expectations about one another, and the actions they take based on those expectations will constantly change.

Finally I point out that due to Brian Arthur and others, a new economic theory is arising based on "increasing returns." Standard economics assumes "diminishing returns." For example, if I add fertilizer to my soil, the first batch increases my crop yields a great deal; but each additional batch brings ever smaller increases in crop production. When economists make mathematical models of markets, decreasing returns helps those models gravitate toward a stable equilibrium. But Arthur and others have pointed out that in high technology sectors such as computer and software design, there are often *increasing* returns. The first version of Microsoft's Word program cost a lot to produce. Copying it is nearly free. The more such programs are copied and sold, the greater the returns. Arthur and others have explained that under these conditions, an inferior technology could get ahead of a superior technology and irreversibly win out. Thus, among the many alternative equilibriums that the economy might reach, the one it actually *does* reach depends upon the detailed history, or pathway, that it has taken. Arthur and others have also shown that the equilibrium the economy reaches may not be the one with the best set of technologies. Increasing returns takes us beyond at least parts of conventional economics, because one hope of earlier economics had been to show that the equilibrium the economy reached was one which was socially optimal. But, if Arthur and others have shown that multiple equilibriums exist, where inferior technologies can win, then there is no guarantee that the equilibrium reached with be socially optimal. Arthur's work and that of his colleagues, implies that even without the profound difficulty of Darwinian preadaptations in the economy, the behavior of the economy may not be predictable due to history dependence, chaotic dynamics, exogenous shocks, or other reasons.

At the heart of all these theories is a fixed set of goods and services known at the start of the story. But this is a vast simplification. The economy keeps evolving new goods and services. Our species, as I remarked, has moved from a few hundred goods and services fifty millennia ago to 10 billion or so today. We therefore need to consider the growth of what I will call the "economic web" via technological and organizational evolution.

THE ECONOMIC WEB AND THE EVOLUTION OF FUTURE WEALTH

Economists employ the concepts of complementary and substitute goods. A hammer and a nail are complements since they are used together to create value. A nail and a screw are, largely, substitutes, since you can usually replace one with the other.

Now imagine points in a large room for all 10 billion goods and services in the global economy. Draw green lines between points that are complements, and red lines between points that are substitutes. The resulting graph is a depiction of the global economic web as it stands today. Fifty thousand years ago the web would have contained a hundred to a thousand points.

Thus, over time, the economic web has expanded into its adjacent possible. Our task is to understand how it expands, and what role the structure of the web itself plays in its own expansion. None of this is known. But it is virtually inconceivable that this evolution, which persistently creates new economic niches and destroys old ones, is not a central part of economic growth.

Here is a first step. Most novel goods and services enter the economy as either complements to or substitutes for existing goods and services. There is no point in inventing the TV channel changer, for instance, until the television is invented and deployed reasonably widely, and there are multiple channels. So the channel changer is a complement to the television. This simple example demonstrates that, as the economic web evolves, it does persistently create new economic niches for new goods and services that fit functionally, hence sensibly, into the existing web. The web begets its own future in ways that we cannot foretell. But, in addition, the television might become of use for other purposes, say long distance banking, as the computer has come to have new uses with respect to word processing, and the engine block that became the chasse. Once that happens *still new, unforeseeable, functional compliments may fit into the still new niches that the existing good, used for a new purpose, affords.* Thus, it is not at all clear, even for existing goods and services, that we can actually prestate all the uses to which they might be put, hence which new economic niches they might create.

In considering the growth of the economic web, therefore, a central question is whether, on average, each new good or service affords less than one, exactly one, or more than one new complementary or substitute possibility in the adjacent possible. If the answer is more than one, then the web (ignoring for the moment investment capital, adoption of new technologies, and other issues) can grow exponentially. That is, if each new good or service affords more than 1.0 new niches for yet further new goods and services, then new economic niches explode exponentially as the economic web grows in diversity. Under these circumstances the very diversity of the web "autocatalytically" drives its own growth into the adjacent possible, affording ever new economic niches—ever new ways of making a living. This point is made even stronger if we bear in mind the new uses which existing goods might come to have, hence the unexpected new niches that might arise as the complements or substitutes for those new uses. In turn this raises the fascinating question of whether the emergence of new uses of existing goods itself depends upon the diversity of already existing goods in the economic web.

There are intuitive grounds for thinking that the number of adjacent possible goods and services that are complements or substitutes may well be more than one per good. When the car was invented, it created the conditions for the oil industry, the gas industry, paved roads, traffic lights, traffic police, bribing traffic police, motels, car washes, fast-food restaurants, and suburbia in what is called a Schumpeterian gale of creative destruction. The destruction side of the story is the extinction of the horse, buggy, saddlery, smithy, and pony express in the United States, as widely used technologies. *The creative parts of the Schumpeterian gale, gas, motels, and suburbia, etc. are all complements to the car. Together they make a kind of autocatalytic, mutually sustaining economic-technological ecosystem of complements that commandeer economic capital resources into that autocatalytic web and can create vast wealth.* All these ways of making a living are largely mutually necessary for one another and they have coevolved together for about a century. Conversely, the hula hoop seems to have few complements or substitutes. It can enter or leave the economic web without creating an avalanche loss of old ways of making a living, or creating new ways of making a living.

The statistical distribution of complements and substitutes of goods and services and their past evolution are empirical questions for economic historians. Perhaps, over the past fifty thousand years, this distribution has changed in some systematic way, and perhaps this change has played a role in the pace of technological evolution. Perhaps the mean number and distribution of substitutes and complements per new good or service have increased as the diversity of the economic web has increased, in part because the number of potential novel uses of existing goods may be positively correlated with the existing diversity of goods, so that the number of niches per good increases with web diversity. If so, the growth of the web is *hyper*exponential, driven by its own diversity. No one knows. Finding out the truth via an analysis of economic history would seem of the deepest interest.

Both anecdotes and good economic data support the rough idea that the diversity of goods and services in an economy drives its growth. In her famous book *The Growth of Cities*, Jane Jacobs notes that in postwar Milan and its hinterland, and in Tokyo and its hinterland, a web of complementary technologies mutually spurred economic growth. Jose Scheinkman and colleagues studied a variety of American cities, normalized industries for total capitalization, and found a positive correlation between economic diversity and growth. Thus, there are some demonstrated grounds to believe that the economic web autocatalytically drives its own growth into the economic adjacent possible, generating ever new economic niches and evolving future wealth.

Another clue that economic diversity aids the innovation of new goods can be found in the story of the Wright brothers' airplane. Their first successful aircraft was a recombination between a modified boat's propeller, an airfoil, a light gas engine, and bicycle wheels. The more goods there are in an economy, the more potential recombinations there are among them. Put an umbrella down the smoke stack of the *Queen Mary* and you get soot in the first-class cabins. Put the same umbrella behind a Cessna and you get an airbrake. Fifty thousand years ago, the recombination possibilities among the few goods and services were few. Now, if we just consider pairs of goods, N goods yields N^2 combinations, any of which has a chance of being useful.

Typically, new complements or substitutes, like the TV channel changer, are also novel functionalities. The function "change the TV channel while remaining on the sofa" did not arise before the TV itself. Like Darwinian preadaptation, much evolution of the economic web involves the emergence of novel functions. Ceaseless creativity confronts us again. And, as noted above, novel uses and purposes for existing goods arise all the time and create still more new economic niches.

This emergence of novel uses arises because there appears to be no lowest-level or "basement" language of simple functionalities from which we could logically derive *a finite set of all such emergent functionalities.*

A simple example of the "basement" language issue and an apparent incapacity to derive a finite set of all emergent functionalities can be seen in the case of what I will call LegoWorld. Think of an enormous pile of simple Lego blocks. Now consider a crane made out of these blocks. The crane, due to its specific organization of structure and processes, like the human heart, has causal properties not found in the blocks themselves. For example, the Lego crane can be used to carry Lego blocks and dump them at a Lego building site. Or it can be used to lift a Lego beam to the top of a Lego house under construction. Now consider a variety of Lego objects, cranes, platforms on wheels, Lego bridges over toy streams, and so on. Now the Lego crane can haul the Lego platform with a load of Lego blocks across the Lego bridge to a Lego house building site across the toy stream! The functionalities that these can attain alone or in combinations do not seem either to be statable in terms of properties of Lego blocks by themselves, nor even to be finitely prestatable at all! Do we really think we can prestate all the uses to which objects made out of Lego blocks might be put? Like the three jawbones of the early fish that evolved into the middle ear bones of your ear, or the parts of the mouse trap displayed in court in Dover, Pennsylvania, *the potential functionalities of Lego cranes and their Lego cousins seem neither reducible to a basement language of all the functionalities of Lego blocks as Lego blocks, nor statable in a finite and effectively describable list of all possible purposes to which those Lego contraptions might be put, alone or in combinations.* Those uses depend not only on the properties of each of the Lego contraptions, alone, but in relation to one another and the rest of the "relevant" world. But we cannot prestate the "relevant world." Relevance depends upon the specific purpose at hand. And we cannot prestate all possible purposes. Yet, for real cranes and contraptions, those uses might be of economic value if they fit sensibly into the ever evolving economic web that creates the ever evolving useful niches, or purposes, for them, like the channel changer fits into, and is useful, with the television set with at least two channels and some abundance of programming—of course in the presence of the couch potato whose utility function gives the economic value to the channel changer. But the evolution of the economic web loops back to its own partial unforeseeability. It is just the new specific purposes to which the new widget is the answer that drove the engineer trying to invent the tractor to see in the rigidity of the engine block new potential functionality as a chasse. But that preadaptation is partially unforeseeable. The partial unforeseeableness of this creation nevertheless fits seamlessly into the functionalities required by the growing economic web and the purposes it serves, as it persistently evolves, in part, into new functionalities that are seen in and afforded by what already exists. As economist Arthur pointed out above, most inventions find uses that were not their initial purpose. The economic web is truly a self-consistent, co-constructing, coevolving whole. So too is civilization.

In the next chapter, I will argue that this persistent inventiveness of the human mind is not algorithmic. There seems to be no effective procedure to derive all the uses of Lego contraptions or real ones, let alone the evolution of cultures and civilizations that also co-construct themselves from their emergent web of purposes, needs, laws, interwoven societal roles, production capacities, and their ethics, and sense of the sacred.

In short, there is a real economic web, but we don't know much about its structure, how it transforms over time, the roles that diverse positions in the web play in successes and risks to the firms occupying them, the way this web grows into the adjacent possible, how it self-co-constructs, and how it spins off new directions of growth in Schumpeterian gales of creative destruction. But clearly, the structure of this web plays a major role in economic activity, economic evolution, hence in economic growth. We must learn to harness this structure, both to create global wealth and to do so in a way that is consistent with a sustainable planet.

In the remainder of this chapter, I will describe an algorithmic model of the growth of the economic web. I present this model, even though I do not believe that the growth of the web is algorithmic, because it may point to at least one kind of science that we can do about the growth of the web, even if the detailed evolution of the web is partially unforeseeable, and even if all the uses to which technological widgets can be put cannot be prestated. We may still find emergent organizational natural laws. The surprising results from this model raise the possibility that the economy advances into its adjacent possible in a self-organized critical way. (I discuss self-organized criticality below in this chapter.) Thus economic evolution may parallel the evolution of the biosphere, which may also evolve in a self-organized critical way, which suggests that perhaps we can draw some statistical conclusions about how coevolving systems in general enter their adjacent possibles. I end the chapter by questioning my assumption that we cannot know the evolution of the economic web beforehand. It is possible (though I strongly doubt it) that agent-based simulations might someday achieve this.

AN ALGORITHMIC MODEL OF THE GROWTH OF THE ECONOMIC WEB

Imagine a diversity of symbol strings such as (100010100). Each year they grow from the soil of France, hence are renewable resources (figure 11.1).

FIGURE 11.1 The outline of France, showing different resources growing from the soil of France. These symbol strings represent the renewable resources—wood, coal, wool, dairy, iron, wheat, and such—with which France is endowed. As the people use the symbol strings to "act" on symbol strings in "economic production functions," new, more complex products can emerge.

Now imagine a grammar table given by a set of pairs of symbol strings, one on the left, (000), and one on the right, (1010). Call such a pair of symbol strings a "rule." The meaning of this grammar table, figure 11.2, is that, if a symbol string occurs in France in which a string from the left-hand side of one of the pairs appears, then that part of the string is removed and replaced

Grammar Table	
1 1 1	0 0 1 0 1
0 0 1 0	1 1 0
0 0	1 0 1 1
1 0 0 1	0 1
1 0 1	0 0 1 0

FIGURE 11.2 A "grammar" table. Instances of the left symbol strings are to be replaced by the corresponding right symbol strings in each pair of symbol strings, in the symbol strings growing from the soil of France in figure 11.1, as symbol strings act on symbol strings in economic production functions specified by the grammar table.

by the corresponding right-hand string. Thus if a symbol string in France such as (100010100) occurred, its substring 000 would be replaced by 1010.

Next, suppose that symbol strings can act on one another like enzymes and substrates. One symbol string will act on another to carry out the substitution mandated by the grammar table. For example, let two symbol strings occur in France, (100010100) and (000101). If one of the strings has a substring that matches one of the right-hand symbol strings in the grammar table, and it finds somewhere in France a second symbol string with the corresponding left-hand string, then the first string cuts out the substring on the second string and substitutes the right-hand symbol string from the grammar table. In the current case, the string (000101) can act on (100010100) and transform it to (1101010100) by substituting a 1010 substring for the 000 substring. This substitution, like an enzyme acting on a substrate, is a model economic production function. A saw acting on a board yields a cut board.

Given these assumptions, we first model the growth of the technologically possible web over time. At each "year," we examine each symbol string in France, compare it to all the other strings, and see if any allowed substitutions creating new symbol string "goods" occur. Then we move on to the next "year." Each year some set of new symbol strings may be generated in this economic adjacent possible, rather like a chemical-reaction graph.

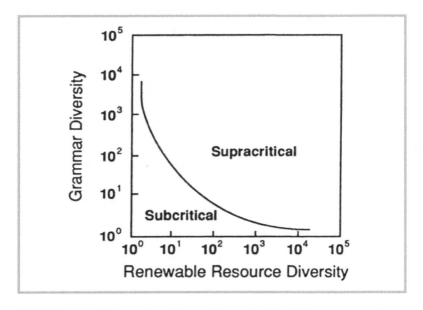

FIGURE 11.3   The number of renewable goods with which an economy is endowed is plotted on the horizontal axis against the number of pairs of symbol strings in the grammar, on the vertical axis. The number of pairs of symbol strings captures the hypothetical "laws of substitutability and complementarity." Production functions link complements. A curve, now proven mathematically, separates a *subcritical* regime below the curve and a *supracritical* regime above the curve. As the diversity of renewable resources or the diversity of rules of the grammar rule table increases, so that the economy lies above the curve, the economy explodes with a diversity of products. The real global economy is almost certainly supracritical, generating an ever changing and typically increasing diversity of novel goods and services. Many regional economies may well be subcritical. In the real technological evolution over the past 50,000 years, the number of rules in the grammar table, figure 11.2, has grown as new goods create new production functions.

Now imagine an *X-Y* coordinate system, figure 11.3. On the *Y* axis put down the number of pairs of symbol string substitution rules in the grammar table. On the *X* axis put the initial diversity of symbol strings in France. Intuitively, two cases arise. If there are very few transformation rules in the grammar table and very few symbol strings in France, then very few or no new strings will be formed. If there is ever a year in which no new strings are formed, the process of generating novel strings stops so the web permanently stops growing. Call this situation *subcritical*. Conversely, if there are enough transformation rules and enough symbol strings initially in France, then new symbol strings will be generated with high probability in the adjacent possible, then be available in the next period for further generation of yet new symbol strings, creating a perhaps sustained explosion of ever novel symbol strings. Call this *supracritical*. Thus, some curve in the X-Y space separates sub- and supracritical behavior. Figure 11.3 shows this curve. It is roughly a hyperbolic curved line that starts where there are many rules and few renewable resources, curves down to the right, and ends where there are few rules and very many renewable resources. This curve is the phase transition between subcritical and supracritical behavior of the model economy. This phase transition has now been shown mathematically with my colleagues Rudi Hanel and Stefan Thurner.

This is a first result that says that there can be a phase transition between an economy that can create only a limited diversity of goods and one that can potentially explode into a diversity of novel goods. The phase transition depends upon the number of technological transformation rules in the grammar table and the number of initial goods. Thus, for example, France alone might be subcritical, Germany alone might be subcritical, but France plus Germany might be supracritical. A tiny country or region with a low diversity of renewable resources might remain subcritical. Its production might never expand beyond a small set of goods. If so, such a country might remain persistently "left behind" by the economic growth of a supracritical global economy in which it could participate only minimally. Economists struggle to understand the persistent disparity between the wealth of nations and their economic growth.

Our model of subcritical and supracritical behavior may be part of the explanation. A very small local economy may remain subcritical. But a wider trading area with a much larger diversity of renewable resources might explode into an ever expanding technologically possible world of ever new goods and services. The United States, the European Union, Japan, Korea, Singapore, Taiwan, and, increasingly, China and India look like supracritical economies. The global economy appears supracritical.

This model leaves economic utility out of the story. It talks only about the technologically possible, not the subset that is also "profitable." A simple economic model, suggested to me by the economist Paul Romer years ago, assumes a single consumer, the king of France, and a utility assigned to each symbol string by some outside "rule." In addition, it assumes that the king would rather have goods today than tomorrow, hence there is a discount function for future goods and services. And it ignores the complexity of markets and assumes a social planner with a planning horizon of some fixed number of years. The planner plans ahead for all the possible goods and services producible over the planning horizon and optimizes the utility of the king of France over the planning horizon, then implements the first-year plan. Some, but not all, of the technologically possible goods and services may be produced. They "make a profit," that is, are worth something to the king of France.

After the first period, the planner advances one period and iterates, planning from then ahead for the planning horizon. Over time, either not much happens and the economy remains subcritical, or increasingly diverse goods and services emerge and perhaps disappear in a supracritical economy.

This model may have important implications. Much of the world remains in poverty. We only partially understand how to drive economic growth, and we have only partially succeeded. The local, regional, national, and global economic web is real, and almost certainly plays a profound role in the emergence of novel economic niches and economic growth. Subcritical and supracritical growth of admittedly algorithmic models of economic webs seems reasonably well established by numerical experiments and analytic mathematical results. Small regional economies may be doomed to limited economic growth because they remain subcritical. Even a place like Alberta in Canada, a rich exporter of oil, timber, and livestock, may fail to generate a highly diverse, ever transforming economic web.

We do not yet know what policy implications may derive from the study of the evolution of the real economic web and its incorporation into economic theory, particularly the theory of economic growth. For example, despite the resistance to tariff barriers in the current free-trade climate, it may turn out that before they open their economics to free trade, small, third-world economic areas should erect tariff barriers and grow local, small, complementary enterprises that build local wealth and production capacity outside a global economy in which they do not have yet have the capital or expertise to compete. Such barriers might even invite foreign investment, since profits could be made in the small local economy. This issue is, of course, familiar in many smaller economies, where lack of tariff barriers has sometimes killed off burgeoning new economic sectors. We need enhanced economic theory to know how best to raise the global standard of living in a sustainable world. Nobel laureate Joseph Steiglitz, a former chief economist of the World Bank, wrote that the bank followed all the theories, known as the Washington Consensus, but did not succeed in turning many countries from "developing" into "developed." Failure to understand the roles of the economic web in economic activities and growth may be part of the reason.

We face limited resources—peak global oil production per day, limited water, increasing population, and a warming planet. There is, at present, unknown practical potential in understanding how the economic web drives its own advance into the adjacent possible, with the persistent creation of ever new economic niches and thus ever new opportunities for future wealth creation. If understanding these processes and how to harness them could help lift the world out of poverty, the practical consequences could be wonderful.

A second mathematical result obtained with my colleagues Stefan Thurner and Rudi Hanel builds a simplified model of the algorithmic economy I sketched above and asks, "If a random good dies, how many other goods die in a Schumpeterian gale of creative destruction?" Conversely, if a new good appears, how many new goods appear in a gale of construction?

The results are quite lovely. Bursts of destruction or extinction of old goods arise with many small bursts or avalanches, and few large ones. Similarly, avalanches of creation of new goods occur, again with many small avalanches and few large avalanches. If you plot the size of an avalanche on the X axis and the number of avalanches of that size on the Y axis of a Cartesian coordinate system, you get a histogram, or distribution of numbers of avalanches of each size. Now the mathematical trick is to transform each axis into a logarithmic scale, so the X axis becomes the logarithm of the size of the avalanche, and the Y axis becomes the logarithm of the number of avalanches of that size. The result is a straight line going down to the right. Such a line is called a power law. Thus the model predicts a power-law distribution of gales of creative destruction or construction.

Creative destruction was made famous by the Austrian economist Joseph Schumpeter, as described above, who pointed out that economic progress entails not only the creation of new goods and services but the extinction of many old ones. As economist Brian Arthur first told me and as described above, when the car comes in, the horse-drawn buggy, saddlery, and pony express go pretty much out of existence. But oil, gas, paved roads, motels, and suburbia emerge. This is a large gale of creative destruction. Fortunes are made by investments in the novel goods and services. The hula hoop created no gale at all. Our model predicts that Schumpterian gales of creative destruction should occur in a power-law distribution whose slope down and to the right in the log-log plot described above is −1.2. A proxy for gales of extinction is the number of firms going out of business in bursts over fixed time intervals. Delightfully, these too are a power law, with a slope of about −1.5. Thus, we seem to predict not only that Schumpeterian gales of creative destruction arise as power laws, but we even seem to get the slope about right. This could be the hint of an organizational law pertaining to the evolution of economies, even if the detailed innovations are not foreseeable. This would be emergent law when the details are unknown.

This model is both interesting and limited. It is, to my knowledge, the first model that predicts the size distribution of Schumpeterian gales of creative destruction and may fit the actual economic facts. That is exciting, of course. But the grammar table of number strings is not the real set of transformations of goods and services. The grammar table evolution is entirely algorithmic. That is, given the grammar table, the initial set of goods and services, and the utility function of the king of France, all possible new goods can be generated algorithmically, and the social planner can calculate algorithmically which subset is produced at each period. In economic reality, however, it seems very likely, as discussed above, that the true evolution of the transformations of goods, or production functions, evolves nonalgorith-mically (I define algorithmic—or mathematically "mechanically"—more carefully in the next chapter), for instance, when tractor manufacturers had the idea of making the engine block serve as a chassis. That is, there appears to be no "mechanical effective procedure," or algorithm, to jump from the chassis and the engine block to using the engine block as the chassis, a novel functionality of the engine block. Nor does there seem to be an effective al-gorithmic procedure to find novel uses of existing goods that create new economic niches. If the true economy is nonalgorithmic, as I suspect, then no algorithmic model will ever capture the details of exactly what goods evolve at what time. Yet the algorithmic grammar model used above sug-gests that it may still be possible to extract statistical organizational laws, such as the size distribution of gales of creative destruction/construction. Science may fail to predict the concrete details, hence not help the CEOs of optic fiber companies that failed when satellite communication displaced them, yet capture major statistical features critical for understanding eco-nomic growth and evolution.

But the full problem is harder. The current algorithmic model avoids prestating all possible goods and services, but it *does* prestate all possible "rules," or production functions in the pairs of symbol strings in the gram-mar table. But this is unrealistic: as we invent new goods and services, we also invent new production functions. No one in the upper Paleolithic fore-saw a machine-tool industry. The machine-tool industry is precisely a new set of economic production functions unknown in the Stone Age. And we constantly use our tools in new ways. So the very set of production functions that drives the growth of the economic web itself grows in nonalgorithmic ways. The present algorithmic model of the non-algorithmic economy thus requires significant maturation before we understand the web's evolution. For example, if the set of grammar rules grows, while the number of symbol strings growing from the ground of France remains constant, the economy may pass through a subcritical to supracritical phase transition. The Indus-trial Revolution may have been just such a transition.

So far we have taken the distribution of failures of companies as a proxy for Schumpeterian gales of creative destruction. This proxy does not measure new firms coming into existence, only those going extinct due to economic change. The observed distribution is a power law. This may hint at a very deep law. The economy, the biosphere, and even the common law, discussed in a later chapter, may evolve into their adjacent possibles in a self-organized critical manner.

In 1988 the physicists Per Bak, Chao Tang, and Kurt Wiesenfeld blasted the physics community awake with their invention of self-organized criticality. I have talked about criticality in earlier chapters as a phase transition or boundary between two states, but now it is time to speak of it as physicists do. If you take a ferromagnet that is not magnetized and tune the external temperature to the critical Curie temperature, it will spontaneously start to magnetize. That is, the magnetic spins in the system will begin to form local clusters where the spins orient in the same direction. At the critical temperature, the size distribution of the clusters is a power law. Power laws mean that there is no natural scale in the system. By contrast, an exponential distribution points to a scale, where the distribution has fallen to $1/e$, where e is the base of the natural logarithms. I should note that demonstration of a power law does not, by itself, prove that the underlying process is due to self-organized criticality (SOC). SOC yields power laws, but so do other processes.

Bak, Teng, and Wiesenfeld invented the "sandpile model." Imagine a table onto which sand is being slowly added at a constant rate. At first, a sand pile will grow and reach the edges of the table. Thereafter, sand slides will begin to avalanche off the sides of the table. If one plots the size distribution of these sand slides, they, too, are a power law. The wonderful feature of this system is that, unlike tuning the temperature to the critical Curie temperature, here sand is added slowly, but no tuning occurs. The system is self-organized critical.

The day at the Santa Fe Institute that I heard Per Bak lecture on self-organized criticality, I remarked to him that the evolutionary record shows small and large "avalanches" of species-extinction events. Maybe the biosphere was self-organized critical. This led Per, Kim Sneppen, Ricard Sole and me in separate efforts to invent models of coevolving species where extinction events could occur and propagate, like the sand avalanches, among species. Sure enough, we produced a spate of models showing that such systems yielded a power-law distribution of extinction events. My own model also produced a power-law distribution of the lifetimes of individual species—about which I will say more in a moment.

The best available evidence on actual extinction events in the evolutionary record best fit a theory in which the size distribution of extinction
events is a power law. Thus the biosphere may be self-organized critical. In
addition, where good data are available, they show that the lifetime distribution of the taxonomic level just above species, genera, is also a power
law. And to come back to the economy for a moment, the lifetime distribution of firms is also a power law.

Now the biosphere is a coevolving system entering the biological adjacent possible. As I have now repeatedly stressed, we have no idea how co-
constructing coevolution achieves systems of such magnificent persisting
complexity as the biosphere with no one in charge, and where Darwinian
preadaptations that are part of this co-construction appear partially lawless. Yet the biosphere constructs itself, evolves, and has persisted for 3.8
billion years. Why has this process been persistent and expanded in
species diversity into its adjacent possible? Do any laws govern this
expansion? We do not understand what I like to call the conditions of
coevolutionary assembly of complex systems. The same ignorance applies
to the evolution of the economy, and as I will discuss in a later chapter, to
the evolution of the English common law. All are cases of coevolutionary
assembly of very complex, interwoven systems with no one in charge.

In my book *Investigations,* I wondered whether there might be general
laws about how self-creating complex systems such as these advance into
their adjacent possible spaces. I framed this as a kind of "fourth law" of thermodynamics for self-constructing open thermodynamic systems like biospheres. Such systems may be self-organized critical. Rather boldly, in
Investigations I hoped for more, that as an average trend, the biosphere might
grow into its adjacent possible in the nonergodic universe in such a way that
the total diversity of organized processes that can happen is maximized.
Now, emboldened by Ulanowitz's results for ecosystems described in chapter
7, I might conjecture that the biosphere evolves so that the product of total
work done times the diversity of work (organized processes involving the
constrained release of energy) done is maximized. But if my fourth law is
true, it may apply more generally to all kinds of co-constructing, co-evolving
systems, including the economy and even the common law. All may be self-
organized critical, and maximize as an average trend some unknown generalization from the physicist's "work," to something like "activity."

In summary, for the economy, the lifetime distribution of firms is, in the best evidence I know, a power law. The size distribution of firm-extinction avalanches is also a power law. This is at least a hint that the economy is self-organized critical, and that such criticality has something deep to do with how economies, the biosphere, and other co-evolving, co-constructing, complex systems advance into their adjacent-possible spaces. The hypothesis may well be false, but the clue seems too big to ignore. Beneath it may lie the principles of coevolutionary assembly. If such laws apply, they cannot be reduced to particle physics, the standard model plus general relativity, or whatever lies deeper in physics. We are far beyond reductionism.

I end this chapter by wondering if I am correct that the evolution of the economy is both unpredictable, like the biosphere, and nonalgorithmic, where algorithmic means, roughly, effectively computable. (I will discuss *algorithmic* in detail in the next chapter.)

It is possible to build computer models of the economic web. Object-based computer languages such as Java can have objects, such as carburetors, that come with "affordances" like "is a," "does a," "needs a," and so forth. Using these, the Java-object carburetor can find that it needs other parts of the engine. At my company BiosGroup, Jim Herriot made the first example of a self-assembling computer model chair. Thus, with effort, we could actually know the current economic web. We would list for each good and service its current complements and substitutes, or list its affordances and use a search engine to "calculate" its complements and substitutes. I actually have an issued patent to accomplish just this, which may only serve to show how far scientists stray from their home training, in my case philosophy and medicine.

The issue is whether it is possible to simulate the evolution of the real economic web. Let's ignore things like access to capital, technology adoption, and utility, and just consider the technologically adjacent possible economy. Might object and agent based models be able to simulate this evolution? I doubt it.

I have already discussed the apparent nonalgorithmic character of LegoWorld. In mathematics, when David Hilbert tried to make mathematics "mechanical" or algorithmic, Kurt Gödel famously showed him that this was not possible. Gödel showed that, given any set of axioms for a sufficiently rich mathematical system, such as arithmetic, there are "true" statements in that system that cannot be derived from the axioms. I suspect the same is true here. In any economic system there will always be novel, emergent functionalities that are not derivable from any basement set of functionalities.

This means that describing the real economic web and its growth is even more difficult than we thought. Beyond LegoWorld, consider the real, but homely screwdriver. We naturally think its complement is a screw, and its substitute is a hammer used with a nail. But think of all the uses to which I might put the screwdriver: I might use it to open a can of paint. I might use it as a more general lever. I might use it to jam open a door. I might use it as a paperweight. I might use it to defend myself against an assailant. I might use it to scrape paint—indeed I have, as I have used a screwdriver to open a can of paint. Probably you have, too. Each of these uses, of course, calls forth its own complements and substitutes. How can we possibly prestate all possible uses of the screwdriver in all possible environments? Did you know that it can be used for opening coconuts when you find yourself cast away on a desert island? With a rock as a hammer, perhaps you can use the screwdriver to cut down small trees to make a dwelling, and then use it to cut small vines to serve as ropes to lash the trees together. I think you get my point.

The novel functionalities of everything, not just fundamental inventions like the transistor, but familiar items like the screwdriver used as a paint-can opener and the engine block used as a tractor's chassis, seem to afford broad potential to flow into the economic adjacent possible. Furthermore, it seems unarguable that these novel functionalities are invented by the human mind. I will present reasons in the next chapter to think that not all actions of the mind are algorithmic.

We seem to confront in technological evolution a ceaseless creativity, because the adjacent possible affords ever new niches, because present goods can be used for novel purposes and hence afford new economic niches, and because via novel functionalities, almost certainly seen and seized upon nonalgorithmically, we invade the adjacent possible. This economic evolution, like that of the biosphere and of human culture and history, is part of the endless creativity in the universe. We have come a long way from reductionism, and a long way from pure science to the inclusion of practical reason, bounded rationality, and whatever else is involved in our economic decisions.

I close this chapter with what is surely true and what is almost certainly true. First, it is surely true that there is an economic web and that it has grown into its adjacent possible; and it is almost certainly true that the structure of the web partially governs its growth. Thus the structure of the web, the ways it affords new economic niches, and the high probability that its diversity is positively correlated with its growth may have the most profound implications for our understanding of economic growth at all levels. Given that much of the world remains in poverty, a better understanding of the roles of the economic web in creating wealth seems imperative.

Second, the analogues of Darwinian preadaptions do occur in the evolution of the economic web. We simply do not know ahead of time what these will be, nor can we make probability statements about them. This limits the applicability of competitive general equilibrium, of game theory, and of rational expectations. And it limits the idealization of the ideally rational economic man.

Third, if we truly cannot say ahead of time, then the way a CEO lives his life and guides his company is a combination of rationality, judgment, intuition, understanding, and invention that goes far beyond the purview of normal science, and far beyond the normal purview of rationality and "knowing." We are called upon to reintegrate our entire humanity in the living of our lives.

Despite these limitations, the simple algorithmic model I presented offers hope that even if we cannot predict what new goods will arise and what old goods will perish, we can find organizational laws that make statistical statements about these processes, and use that knowledge in our economic planning. In short, even if the evolution of the economy is nonalgorithmic and not predictable in detail, algorithmic models of its growth may still capture some useful features and reveal new emergent laws. Among these is that the economy may coevolve into its adjacent possible in a self-organized critical way, which in turn may hint at principles of coevolutionary assembly. If these latter points can be shown to be true, we will have understood something very deep.

Like the biosphere, the economy is ceaselessly creative. Beyond the reach of reductionism, this human creativity is sacred.

12

MIND

Mind and consciousness are central to this book. Consciousness is the aspect of our humanity that is most obviously—and famously—incompatible with reductionism; it is seen by many as our best evidence, indeed for some a proof, of the existence of a Creator God. If there is a soul, we tend to think it is located in consciousness or synonymous with it. Thus when the Catholic Church recently embraced the theory of evolution, it was careful to restrict its endorsement to the evolution of the body, not the mind. We all have different interpretations of what consciousness is.

In this chapter, I will describe what mind is not. In the next chapter, I will discuss consciousness. I believe the human mind is not algorithmic, and is not a mere "computational machine." Rather, I believe that the mind is *more* than a computational machine. Embodied in us, the human mind is a *meaning and doing organic system*. How the mind is able to generate the array of meanings and doings it does is beyond current theory.

Algorithms

It is not clear that we have a theory of the human mind, let alone an adequate theory showing that the mind is algorithmic. Still, many cognitive scientists and neurobiologists believe the mind is always algorithmic. Before we begin to address this point, what is an algorithm? The quick definition is *an effective procedure to calculate a result*. A computer program is an algorithm, and so is long division.

The mathematical basis for the modern concept of an algorithm was laid down in the early twentieth century. The mathematician Alan Turing asked himself, in effect, what he did when he computed a result. He noted that he wrote symbols on a page of paper, modified the symbols according to a precise set of rules (the rules of mathematics), wrote new symbols as a function of that stage of the calculation, and carried on until, if all went well, he had the answer he wanted. Turing abstracted these procedures into a machine. This remarkable device has a number of internal states given by symbols. In addition the machine has an infinite tape divided into squares, on each of which one symbol or no symbol is written, and a reading head, with the internal states, that reads the symbols. The reading head reads the symbol at a specific starting point on the tape, which tells it to remain in position or move one step to the left or right, and also tells it whether to write a new symbol on the tape at the current location, depending upon the symbol it reads and its internal state, also given by a symbol. In addition to writing or not writing and moving or not moving the tape, the reading head might shift the machine to a new internal state. The process repeats. Turing showed that this device, which came to be called a Turing machine, could perform any "effectively computable" computation. Indeed, we now define "effectively computable" to mean computable on a Turing machine. Interestingly, not all things mathematical are effectively computable. Turning himself showed that most of the irrational numbers are not effectively computable.

The Nobel laureate Herbert Simon correctly points out that a computer program is what is called a "difference" equation. A difference equation is a cousin to the differential equations of Newton, where time varies continuously. In a difference equation, time does not flow continuously but is broken into discrete units. The difference equation uses the current state of the computer system and its current inputs to compute its next state and its outputs. Simon's view fits with two streams of cognitive and neuroscience, discussed next, in which the mind is a computational information processing system.

Turing's triumph ushered in much of the past half century of cognitive science and neuroscience, for it became the deep theoretical framework of the now widely accepted view that the mind is a computational machine of some sort. Shortly after Turing, the mathematician John von Neumann invented the architecture of the contemporary computer. The way was open to view the mind as a computer of some kind. In this view, the mind is algorithmic. Another groundbreaking early piece of work by Warren McCulloch and Walter Pitts in 1943 abstracted the properties of real neurons and treated neurons as "logical devices" exactly like my Boolean genes discussed in chapter 8. These logical neurons could be on or off, 1 or 0. In one interpretation, the on/off states of the logical neurons "stood for" or "meant" the truth or falseness of simple propositions, called atomic propositions, such as "red here" or "A flat now." McCulloch and Pitts showed that a network among such formal neurons called a "feed-forward" network—that is, one in which there were no loops in the connections among the neurons—in principle could calculate any complex logical combination of these atomic propositions. So, they concluded, the mind could be a computer manipulating "information," here interpreted as "logical atomic propositions" encoded by the on/off states of formal neurons.

Two major strands have unfolded from these early beginnings over the intervening fifty or more years. The first has been dominated by attempts to understand symbol processing by the human mind, such as that in human language and grammar usage, human logical deductive reasoning, and automated mathematical proof algorithms meant to show that computers, like mathematicians, could prove theorems. A flood of algorithms along these lines has been produced and studied.

The second strand is related to British nineteenth-century associationism, namely the association of ideas. The results have been powerful. A very simple introduction to these ideas, now called connectionism, has already been presented in chapter 8. We need only borrow much the same ideas you have already seen. Recall from chapter 8 my Boolean network models of genetic regulatory networks. In chapter 8, the binary, on/off, 1/0 variables were considered active or inactive genes. Now let us just rename the genes as neurons, and my logical genes become McCulloch and Pitts's logical neurons. There are two differences. First McCulloch and Pitts limited themselves to networks with no feedback loops. But the real brain and real genetic regulatory networks are replete with feedback loops, so we will utilize such networks. Second, a minor mathematical detail. The simplest logical abstraction of neurons are that some inputs to a neuron inhibit it from firing, others excite it. The corresponding Boolean rules that match these requirements for a neuron with, say, K inputs are a special "linearly separable" subset of all possible Boolean rules for a logical element with K inputs. My networks in chapter 8 happen to use all possible rules; Boolean mathematical models of neural nets do not. From chapter 8, recall that the synchronous Boolean networks have states, each the current values of all the genes or neurons in the genetic or neural network, and that each state flows in the synchronous case to a unique successor state. Thus the state space of all the possible states in the system consists in *trajectories* of states that flow through one another in sequence and enter into one *state-cycle attractor*, where that attractor and the states flowing into it are the *basin of attraction* of that attractor. More, the state space may have *multiple attractor state cycles*, each draining its own basin of attraction. The state transitions in state space are the analogue of the connection of ideas. This picture suffices to give the simplest picture of the connectionist framework for neural networks. The word *connectionist* simply refers to the fact that such networks are interconnected nodes, here representing neurons.

In the genetic network framework of chapter 8, you will recall that a state-cycle attractor was interpreted to be a distinct cell type. The different attractors were interpreted to be the different cell types in multicelled organisms like us. Each cell type corresponds, on this model, to different combinations of on-and-off activities of all the (N) genes in the network. Here in the neural network framework, in the simplest interpretation, the on/off state of each neuron "means," or "stands for," the truth or falseness of an atomic proposition, again, "A flat now" or "red here." A state of the network is then a combination of true and false statements about the N different atomic propositions represented by the N neurons in the connectionist neural network. Then a state-cycle attractor might stand, for example, for a *concept or a memory of the logical combinations of the N true and false atomic propositions encoded in its states.* The states lying in the basin of attraction of that attractor that flow to it are the "generalization" of the concept, for example, all things co-classified as frogs, where some "canonical" frog is encoded by the attractor itself. Again, to be specific, suppose that each logical neuron again stands for—means—a logical atomic proposition. The on/off state of that logical neuron is again the truth or falseness of that logical atomic proposition. Then the attractor is some logical combination, or combinations, of those propositions that jointly mean "frog." States in the same "frog" basin of attraction are combinations of true and false atomic propositions among the N neurons that encode froglike entities. Thus the basin of attraction is the generalization class of the concept "frog" encoded in the attractor. All froglike entities are classified together by the network as frogs. Different attractors encode different concepts, frog, table, skates, or they encode different memories.

Contemporary mathematical connectionist models are more subtle than this simple, synchronous Boolean network and use model neurons that have continuously varying levels of neural activities, or more complex noisy-stochastic levels of neuron-firing activities, as discussed below. But the central idea remains the same. In its *neurobiological* interpretation, the firing of real neurons, discussed in the next chapter, computes and yields a computational processing of information. Again, attractors that exist in the dynamics of the neural network can be interpreted as concepts, or classes, or memories, or more broadly. Next, suppose atoms of conscious experience are produced by, or are identical with, the firing of specific different neurons or clusters of neurons, where these different atoms of experience are the "neural code" embodied by those specific neurons or clusters of neurons. Then the real neurons process this experiential information to yield mental experiences, including classification and memory of experiences.

Within the computational theory models of the brain, the two strands do not fit easily together. The symbol-processing first strand, easily able to capture some of linguistic grammar and production of logical proofs, does not readily carry out the pattern recognition, for example, of frogs, and pattern generalization, for example to froglike entitites, that is natural to the connectionist view. Conversely, the connectionist picture of basins of attraction and attractors has a difficult time accommodating the symbol processing properties of the first computational strand.

I will discuss below whether the first strand, the symbolic strand, or the connectionist strand seem adequate to account for mental activity. I stress again that most cognitive scientists and neurobiologists today adhere to some version of the connectionist computational model of the brain as an information processing system.

SOME NEUROSCIENCE CONSISTENT
WITH A COMPUTATIONAL MIND

Neuroscientists have spent much of the past thirty years accomplishing things we would have thought impossible, on the basis of the hypothesis that the mind is an algorithmic, complex computational system.

Perhaps one of the earliest results in this direction was the discovery of what are called receptor fields. By moving a tiny light source across the retina of anesthetized cats and recording from retinal neurons, researchers found that some neurons had a circular or oval receptor field, a small region on the retina with a modest number of the cells, rods, or cones, that respond to light. Typically, when the light hit the center of the receptor field, the neuron increased its firing rate. If the light hit the outer edges of the receptor field, the neuron lowered its firing rate. Such a receptor field is called an on-center, off-surround field. Other fields are off center, on surround. Remarkably, the on-center can be bar-shaped. The bar responds optimally to a tiny rectangular light beam shining parallel to the bar, but less strongly to a light shining perpendicular to the bar. The same field is suppressed if the light shines on the outer margins of the receptor field. These receptor fields are called edge detectors. Elegant work has mapped edge detectors of each orientation to the visual cortex of the brain and shown that neurons with any given orientation tend to form connected lines of adjacent neurons across the surface of the cortex. Lines corresponding to all possible orientations can merge at single points on the cortex in what are called singularities. These kinds of elegant results clearly suggest that part of what creates visual experiences is a kind of *neural calculation* driven by sets of neurons controlled by edge detectors.

One notable kind of visual experience is geometric hallucinations. The mathematician and theoretical neuroscientist Jack Cowan has carried out detailed work to explain these. Cowan and others showed that there is a specific mapping of the visual field onto the visual cortex that happens to be a logarithmic spiral pattern. Using mathematical models of waves of neural excitation propagating across the visual cortex, he has shown how, given the mapping of the visual field to the visual cortex, these waves may form weblike figures or geometric patterns that correspond to specific hallucinations various people have reported. Cowan asserts that the common near-death experience of a tunnel with a light at the end is probably due to a simple neural activity wave process propagating across the cortex. (I had rather hoped the light really was an entrance to eternity.)

There is also evidence that our neurons may make decisions before we do. It has been found that before you move your left index finger, specific neurons appear to fire at an enhanced rate; this may be the neural underpinning of your intention to move your finger. The puzzle is that the neurons appear to fire *before* you experience the intention. If so, how does the intention you experience "cause" you to move your finger? Perhaps it is the firing of the intention neurons that causes your finger to move, not what you experienced as a feeling of making the decision. No one knows.

Findings like these reinforce a general belief among neurobiologists that the brain is a physically classical (not quantum) system, and that the mind is algorithmic and an information computational network. But the mind may be a classically causal system, despite philosopher of mind Donald Davidson's idea of anomalous causation where no law is to be expected relating neural behavior and all cognitive experiences, and yet not be algorithmic at all. Not all deterministic dynamical systems that change in time are computing algorithms. For example, the planets of the solar system move according to the laws of general relativity, deforming space by their mass and following shortest pathways, called geodesics, in that deforming space. The planets are not computing their trajectories. Nor does a windmill compute its trajectory. Thus we can accept the wonderful results of the neuroscientists, accept that the mind, via neural behavior, is classically causal, and refuse the conclusion that the mind is computing an algorithm. I will discuss grounds next to lend doubt to the claim that the mind is always computing an algorithm.

COGNITIVE SCIENCE, THE ALGORITHMIC MIND, AND ITS TRAVAILS

Cognitive scientists are the other important group who hold that the mind is algorithmic and a computing system. Like the neuroscientists, they, too, have accomplished remarkable feats, such as showing that in defined problem settings, with defined objects having defined capacities, algorithmic procedures can solve specific problems. A robot moving to avoid objects in a room and finding a source of electricity for food is an example. But these successes are actually quite problematic.

One initial clue that the mind may not be algorithmic is the famous incompleteness theorem of Kurt Gödel. In 1931, Gödel published a proof showing that *mathematics itself cannot be made algorithmic*. For any sufficiently rich system of axioms, such as arithmetic, there will always be mathematical statements that are *true*, given those axioms, but *not deducible from the axioms* themselves. These statements are called *formally undecidable*. Yet mathematics itself grows, and many mathematicians believe that the way mathematics grows and diversifies is *not itself algorithmic*. One example might be the discovery of non-Euclidian geometry in the nineteenth century by a German mathematician, Bernhard Riemann. Here the parallel axiom of Euclid, that parallel lines never cross, was negated. With this new axiom, *which could not itself have been derived deductively*, the result was the generation of an entire new branch of non-Euclidian geometry such as that underlying general relativity. The insight, intuition, and modes of thought that led Riemann to think that negating the parallel axiom might be deeply interesting and lead to important new mathematics are not obviously algorithmic. Perhaps an even more striking example is the invention of an entire new field of mathematics called topology by the mathematician Leonard Euler. He was famously confronted by the following intriguing problem concerning the German city of Konigsberg. The city is on a river with an island in the middle. Seven bridges connect the parts of the city on the shores of the river with the part of the city on the island. The question arose as to whether one could walk across the bridges, each in a single direction and each only once, starting at and returning to any single point in the city. Euler brilliantly realized that the detailed geometry of the city did not matter, only the different internally connected regions of the city, the two parts on the two shores, the part on the island, and the connections between them due to the seven bridges. Euler proved that no one could carry out the walking task described above. In doing so, Euler invented an entirely new branch of mathematics, a "geometry" that does not depend on the details of shapes, that is, topology. What Euler *did not do* in inventing topology was to deduce it from existing mathematics. Why should we think his invention was algorithmic? By what algorithm could or did he invent topology?

But the view that the mind must be algorithmic faces more problems. A characteristic example of how the arguments that the mind is algorithmic become problematic is the concept of categorization. We categorize all the time—for instance, we lump robins and penguins together as birds. Douglas Medin's research into human categorization has shown that we do not really know *how* we form these categories. The classical idea, dating from Plato, is that members of a category all share one or more "essential" features. It turns out that this is false. Wittgenstein was perhaps the first to point out that members of categories might have no feature in common. "Games," for instance, need not share any single property. Rather, he argued, they share a family resemblance. But there is no one property that is common to all games. This observation led to a "probabilistic" theory of category formation, in which ranges of properties are more or less central to category membership. Even this turns out not to work all the time.

Our naïve notion of category rests on a concept of similarity. Robins and penguins are similar because they share a number of features. But there are severe problems with the concept of similarity itself. What makes the features of two different things similar, even if we agree beforehand on the relevant features of the objects to be compared? How do we define it? It has turned out to be extremely difficult to clarify this seemingly simple idea. And the problems are even worse. How do we pick out the *relevant properties* that should be considered similar? Any two objects have a vast set of properties. A crowbar and a robin are similar in weighing less than a hundred pounds, being on the Earth, and being 239,000 miles from the moon. It is not entirely obvious just which features, out of unlimited possibilities, we should choose as relevant for assigning objects into categories.

As Medin points out, you could place babies, photo albums, personal papers, and money into a single category, because they are what one might remove if one's house were on fire. This leads Medin to think that categories themselves are based on *theories*, but that some sense of similarity is still requisite for categorization. But this idea still falls far short of explaining human categorization as an algorithmic activity. Where do Medin's *theories* come from? Are they, too, derived algorithmically? Was Riemann's new axiom derived algorithmically? Or Euler's invention of topology? The story is not helped if we use a *connectionist* model of categorization, with its basins of attraction and attractors, for again where do Medin's theories that underlie categorization, hence must alter the basins of attraction so appropriate things are coclassified as similar, come from? Are these theories, too, either algorithmic or connectionist productions? What convinces us that the invention of such theories underlying categorization must be algorithmic? Again, Reimann's innovative move to deny the parallel axiom of Euclid, which led to his discovery of non-Euclidian geometry, was not obviously algorithmic, nor was the discovery of topology.

Yet such categorization is required to "state" a problem setting and a problem "space," and to seek a solution to the problem in the prestated problem space. Take, for example, our robot navigating among predefined obstacles and solving the problem of finding a prestated power source to "feed" itself. For the builder and programmer of the robot to solve this problem in the problem setting, it is essential that the *relevant features* of the objects be *prestated*. Yet as we have seen, prestating a problem space is fraught with difficulty. Seeing all the potentially relevant properties of the robot with respect to all the potentially relevant features of the room that might arise as part of the solution to some still unstated problem cannot be finitely prestated. Nor does it appear that there is an algorithm that is an effective procedure to generate all those potentially relevant relationships. How would such an algorithm "work"?

To take another example, the engineer who noted that a tractor's engine block could serve as a chassis clearly stepped out of the familiar problem space to find a novel functionality of the engine block and a novel solution to his problem. The rigidity that made the block useful for this purpose was *not* prestated as a *relevant feature* of the engine block in the problem setting. This example shows how extremely hard the "frame" problem is. For cognitive scientists using computer models of human mental problem solving, how is the frame of the situation prescribed? Roughly, the frame is a list of the relevant features of the situation. No one knows what to do about the limitations on problem solving that result once the relevant features are prespecified. The rigidity of the engine block was not picked out as a relevant feature of the engine block for the engineering purpose of building a tractor. Thus no algorithm operating on the prestated relevant features of the engine block could "find" the solution that the block's rigidity meant it could be used as a chassis. Algorithmic problem solving outside of those prespecified relevant features appears precluded. Yet we do it all the time. At its heart, this is part of a radical conclusion: the mind is not (always) algorithmic.

While we can write workable computer programs with bounded frames and carry out algorithmic problem solving within the prestated set of possibilities, say of the relevant capacities or "affordances" (such as "is a," "has a," "does a," "needs a") of the objects, in the prestated problem situation, it is not at all clear that human minds are similarly limited. Indeed, it seems clear that we are not. The rigidity of the engine block is an "affordance" that was not in the initial frame of the problem setting of the engineers. The absence of "frame" limitation by real people solving real problems, tractors and otherwise, suggests that the human mind is not algorithmic.

In the last chapter we considered our incapacity to produce a finite list of uses of Lego contraptions. This strongly suggests that our finding uses for Lego contraptions is not algorithmic. And consider again from the last chapter the humble screwdriver. Of course its normal function is to screw in a screw. But how many other novel uses can the screwdriver be put to? It can be used to open a can of paint, used as a more general lever, used to scrape putty from a frozen window, used to defend yourself against an assailant, used as an object of art, used as a paperweight. The screwdriver can be used to carve your initials on a fine tabletop, spear a fish, crack a coconut, chop down a tree using a rock to hammer it if you are on an isolated island making a hut. There appears to be no prestatable bound, or frame, that limits the uses to which we might put the screwdriver in diverse circumstances. There appears to be no finite list of affordances of the screwdriver. Do we think we can prestate all possible tasks in all possible environments and problem situations, such that we could construct a bounded frame for screwdrivers? Do we think we could write an algorithm, an effective procedure, to generate a possibly infinite list of all possible uses of screwdrivers in all possible circumstances, some of which do not yet exist? I don't think we could get started. And if we did prestate some finite list of tasks that we could perform with a screwdriver, the prestated relevant features would limit the problem solving that we could then carry out. So while we can, in algorithmic fact, construct models in which problem solving occurs in a bounded space of tasks and "affordances," in reality, there appear to be no such bounds. If you don't believe me, watch Bond movies as the remarkable 007 ingeniously uses whatever lies at hand in his never-ending crusade to keep Great Britain safe. And MacGyver is our other innovative hero.

Cognitive scientists have done remarkable work in describing ways to carry out categorization, representation of problems, and solution to problems in defined spaces. And they can write wonderful, workable computer programs following their own prescriptions. Herbert Simon would say that such models actually reflect how the human mind works. I admire Simon enormously, but disagree sharply. At every step there arise profound problems: with categorizations themselves, with the use of these to achieve a representation of a problem, with the bounds placed on the affordances or capacities of the entities to be used to solve the problem, and so forth. The human mind, like a ghost ship, keeps slipping free of its computational moorings to sail where it will. It does so because it is nonalgorithmic. This freedom is part of the creativity in the universe. It is our own creativity as humans.

The chapter on economics pointed out that inventions can involve recognizing the usefulness of a feature of an object or situation for a new purpose, perhaps a purpose never before seen in the history of the universe. Here is another reason for thinking that we cannot prestate all possible functionalities or purposes of human actions: Suppose I cash a check. This is a common action, yet it could not have taken place fifty thousand years ago. Check cashing required the social invention of money, credit, banks, contractual relations involved in banking money and cashing checks to retrieve that money, legal systems to enforce such contracts, governments capable of creating and sustaining those legal systems, and so on. My act of cashing a check occurs within a social context of cultural invention over centuries. What algorithmic process, fifty thousand years ago, could have listed all of the actions I can now undertake? The social institutions as well as the objects had to be invented. Many of those social institutions, like novel uses of the screwdriver, were not algorithmic derivations. They were creative inventions. The entire cultural evolution that allows me to cash a check is a further case of historical, contingent, innovative, ceaselessly novel emergence. Like the biosphere in which partially lawless Darwinian preadaptations arise and create new functionalities that in turn afford novel niches within the same organism or other organisms for still further adaptations or Darwinian preadaptations that "fit" together with the initial preadaption, and like the evolution of the economic web where the channel changer fits the television with several channels and the couch potato, cultural evolution is also co-constructing, and ever coevolving with itself and other cultures. It is, again, the creativity in the universe, here carried out by human beings. Could we predict this evolution?

There are further grounds to question that the mind is always performing an algorithm or a computation by appeal to the failure of eliminative reductionism. Consider the statement, "Kauffman was found guilty of murder in Santa Fe, New Mexico, on July 7, 2006." What is required to understand this legal language? Clearly we need to understand the concepts of guilt, innocence, evidence, law, legal responsibility, and a whole host of other interlinked concepts that Wittgenstein called a language game. His central point is that we cannot eliminate this legal language as if it were a mere "shorthand" that could be replaced without alteration of truth status by a set of statements about the ordinary actions of human beings that can be prestated as a finite set of necessary and sufficient conditions about the truth of the legal statements. Again, eliminative reductionism fails. But this implies that we cannot deduce legal talk from language about the ordinary actions of people, let alone from the truth of statements about quarks and gluons. So legal language, and thus our entire edifice of law, is emergent, a child of the evolution of social culture, not reducible to language about the acts of individuals, nor to particles in motion.

Nevertheless, we learn legal language. Assuming Wittgenstein is correct (virtually all philosophers agree that he is), then we cannot learn legal language algorithmically based on language about normal human actions, nor from descriptions of quarks and gluons. Not only is legal language an example of emergence in the social sphere, the fact that we learn it is an example of the human mind behaving in a nonalgorithmic manner. Since we cannot deduce legal language from talk about ordinary human acts, we cannot learn legal language algorithmically from talk about ordinary human acts. Then the mind need not be algorithmic.

The same Wittgensteinian point about language games that cannot be reduced to other language games seems to apply equally to the connectionist model. If the neural encodings of formal or real neurons *mean* or *stand for* aspects of *ordinary human action* in this example, *there is no way to jump to legal language*, and our understanding of guilt, innocence, evidence, legally admissible evidence, legal responsibility, and the family language of concepts that are our legal understanding. Yet we do understand legal language. In short, for the connectionist model to work, the *encodings* of single neurons or groups of neurons of experiences and the context and situated meanings and understandings of experiences seem to have to shift, or change, or climb to higher levels, say, from body positions and vocalizations to ordinary intended actions and statements to legal responsibility for action, in ways that are entirely unclear on a logical computational model. In short, while we might want to accept a view in which the activity of specific neurons or groups of neurons were the *neural correlates of specific aspects of experience* it is not at all clear how logical computations on these units of experience come to be our understanding of say, all the meanings of Lady

Macbeth exclaiming, "Out, out damned spot," within its contexts of power, murder, law, justice, and feared punishment.

Or again, recall the failure of eliminative reductionism in our understanding the statement, "England goes to war with Germany." As we saw, there is no prespecifiable finite set of statements about the people involved that are jointly necessary and sufficient for England to manage to go to war with Germany. The "computation" concerning the actions of people such that war ensues cannot be carried out because the conditions for the truth of the statement that war breaks out between England and Germany are not prespecifiable, hence the statement that England goes to war with Germany is not algorithmically derivable from claims about ordinary individual human actions, let alone quarks and gluons. Yet we understand the statement.

I give a final amusing example. Years ago I sat in front of our living-room coffee table on which my computer rested, a cord plugged into the floor socket. I feared my family would trip on the cord and pull my computer off the table, breaking it. I faced a severe problem, as you can plainly see. I now describe the table. It had three broad wooden boards as the tabletop, four fat, round legs, horizontal stringers among the legs, was painted red, and had innumerable chips and all the distances among all the possible positions on the chips. The tabletop's middle board had two cracks, one a half inch at its base, winding a foot across the board, the second somewhat narrower and roughly parallel to the first. The table was seven feet from the fireplace, fifteen feet from the nearest point in the kitchen, twenty-two feet from the stove, 239,000 miles from the moon. . . . I suspect that you are beginning to get the sense that there are an indefinite set of features of my humble coffee table. In fact, I solved my severe problem. I jammed the cord into the wider crack in the middle board on the tabletop, wedged it firmly up into the crack, pulled tight the portion of the cord running from the floor plug to the crack, and felt deeply proud. I felt almost as proud as I do of my machine to water my beanfield, described earlier. This again is the frame problem. Is there an algorithmic way to bound the frame of the features of my table, computer, cord, plug, and the rest of the universe, such that I could algorithmically find a solution to my problem? No. But solve it I did!

The Mind Is a
Meaning-Doing System

A central failure of the "mind as a computational system" theory is that computations, per se, are devoid of meaning. They are purely syntactic. They are devoid of semantics. There are two ways to begin to see this. First, consider the familiar computer and its actual physical structure with memory registers and so forth. As the computer, given its program and input data by *us human users*, crunches its bits, the bits, the 1 and 0 values in the registers, *mean nothing at all*. The bits are just bits, 1 and 0 values carried by some electronic state of a silicon chip, or a water bowl for that matter. Even this statement is an overstatement. The bowl example points out that it is *us humans who give meanings of 1 and 0 to water above or below some threshold level in the bowl*, or one electronic state or another in the register. As physical systems, the water bowl and silicon chip are simply in physical states. *Meaning* was shepherded into our discussion above in discussing the McCulloch-Pitts network of formal neurons, when I said that a formal neuron, on or off, *meant the truth or falseness of an atomic logical proposition such as "A flat now."* That meaning is not inherently located in the on/off state of the formal neuron. The same problem arises in the connectionist view, for again, it is we who invest the formal neurons, on or off, with meanings.

Another way of seeing this is to consider briefly Claude Shannon's famous information theory. Shannon was a telecommunication engineer concerned with passing signals down a noisy communication channel. He simplified a signal to its minimal case, passing 1 and 0 symbols down the channel. His mathematical formulation of information, the negative of the mathematical formulation from statistical mechanics of entropy, quantitates the *amount of information* communicated down the channel. But Shannon carefully never says what information itself is. This conundrum is left, by Shannon, to the *"receiver"* of the information who interprets that information, that is the receiver gives the information its meaning, its semantics. Shannon's theory is purely syntactic, an arrangement of symbol strings chosen out of some set or ensemble of symbol strings. Similarly, the formal grammars derived by linguist Noam Chomsky from Turing's work are purely syntactic, devoid of meanings, devoid, in short, of the semantics of the symbols.

Where, then, do meanings come from? I have discussed what I believe to be the start of an answer. *Meaning derives from agency*. Recall the discussion of the minimal autonomous molecular agent, reproducing, doing at least one work cycle, with a receptor for food and for poison, and able to move toward food and away from poison. We can substitute a bacterium swimming up a glucose gradient for food as our example. Then, I claimed, an increased rate of glucose molecules detected by a glucose receptor as the bacterium swims or orients up the gradient was a *sign* of more glucose up the glucose gradient, and that sign was *interpreted* by the bacterium by *its oriented motion up the glucose gradient*. In the C. S. Peircian sense, the glucose is given meaning to the bacterium by the bacterium's reception of the sign, the glucose, and in its doings, here, swimming up the glucose gradient. The bacterium itself is the receiver. And in this case it is natural selection that has assembled the molecular systems able to accomplish this.

Without agency, as far as I can tell, there can be no meaning. It is a very long distance to human agency and meaning. But it is we humans who *use* the computer to solve *our* problems. It is we who invest meanings in the physical states of the water bowls or electronic states of the silicon chip. This meaning is the semantics missing in the Turing machine's computations. Without the semantics, the Turing machine is merely a set of physical states of marks on paper, or levels of water in a water bowl or electronic states on that silicon chip. Similarly, it is not a wonder than Shannon brilliantly *ignored semantics* to arrive at his quantitative theory of the amount of information carried down a channel. That is why Shannon tells us the *amount of information* passing down a channel, a syntactic quantity, but does not tell us what information *is*.

The mind makes meanings. It makes understandings. We do not yet know very well how it does so. This failure, at least to date, is captured by Wittgensteinian language games, as discussed above. If we cannot logically derive legal language from language about ordinary human actions, yet we can and do regularly come to understand legal meanings, how do we do so? In truth, we do not know how we do so. But we do. Somehow, we are shown specific examples and "grasp" the *legal* meanings.

Thus, while the human mind, central to our human embodied agency, is sometimes algorithmic and sometimes computes, it does some things we do not yet understand; it makes meanings. It sees a potential chassis in an engine block. It sees a new branch of mathematics in the bridges of Konigsberg. So the syntactic, algorithmic, connectionist theory of mind is part of the truth, but it is far from adequate. More, even if we take the step of identifying the firings of specific neurons with specific experiences, as discussed in the next chapter, how we integrated humans make sense of and understand the meanings of those experiences is still deeply mysterious. How indeed do we understand Lady Macbeth's exclamation, "Out, out damn spot," in all its levels of meaning?

The last case I consider concerns the efforts to invent a theory of quantum gravity. The current state of affairs is well described in Lee Smolin's book *The Trouble with Physics*. The cornerstone theories of twentieth century physics, quantum theory (which describes the three subatomic forces of fundamental physics) and general relativity (which describes the fourth force, gravitation), do not fit together. Physicists would like to fit them together in a theory of quantum gravity. Part of the problem is that quantum mechanics is a "linear" theory. The Schrödinger equation is a linear wave equation, allowing the famous superposition of possibilities now being exploited in quantum computers. "Superposition" of linear wave equations means that if a wave of shape A can propagate and one of shape B can propagate via the linear wave equation, then the sum or difference of the two waves, A + B or A − B, can also propagate via the wave equation. General relativity, on the other hand, is inherently nonlinear. Masses curve spacetime, and the curvature of spacetime, constituting gravity, alters how masses move.

One attempt to invent a theory of quantum gravity is string theory. Instead of considering particles as zero-dimensional points, the idea is that there are open or closed strings whose modes of vibration correspond to particles. It is a lovely idea. But as Smolin points out, no one has yet written down string theory as some set of equations. Worse, it now appears that there are something like 10^{500} alternative string theories. Other possibilities for a theory of quantum gravity are being explored, such as loop quantum gravity, by Smolin and others. The central issue here is the following: it is not clear whether quantum theory itself needs to be modified or even derived from some new theory. It is not clear if general relativity needs to be derived from some unknown deeper theory that would yield both it and quantum mechanics. If no deeper theory is needed, then it is not clear how the two are to be united. In short, even the formulation of the problem, the problem statement, remains unknown. Are these two theories to be modified and united? Are they to remain unmodified and united? The idea of replacing particles with vibrating strings was an entirely unknown concept forty years ago. It seems utterly impossible that this terribly difficult, courageous scientific effort is algorithmic. There is no clear set of building blocks, no clear set of desired theorems, no framework for what a theory that derives quantum mechanics and general relativity could be.

Then there are the puzzles of dark matter in the universe, the mysterious substance that makes the outsides of galaxies rotate too rapidly for either Newtonian gravity or general relativity, and worse, the now famous dark energy that is associated with the accelerating expansion of the universe. Do these mysteries provide clues to an eventual theory of quantum gravity? Are the heroic efforts to understand these algorithmic? Suppose string theory turns out to be the ultimate answer, despite the 10^{500} versions currently envisioned. Think of the founding ideas that have been invented and the stunning mathematics that has been invented to arrive at current string theory. It seems impossible that these ideas can be arrived at algorithmically.

Thus I am persuaded on multiple grounds that the human mind is not always algorithmic, nor merely computational in the syntactic sense of computational. The minds of the physicists seeking to unite quantum theory and general relativity are certainly making meanings, certainly inventing.

We are left with major puzzles about the mind and how it works, even before we consider that astonishing enigma, consciousness. Among these are our capacity to see a chassis in an engine block, learn legal language, and seize and act upon a novel opportunity. If the mind is produced by or "identical with" the physically classical activity of neurons and perhaps other cells in the brain, and that system is a classical dynamical system, whether deterministic or noisy and stochastic, it is far from clear at this point how the mind does see a chassis in the engine block, learn legal language, or seize and act on novel opportunities outside of any prestated frame. My sense is that neuroscientists and cognitive scientists do not now know the answers to these deep problems.

But must conscious mind be classical, rather than quantum or a mixture of quantum and classical? Could consciousness be a very special poised state between quantum coherence and decoherence to classicity by which "immaterial, nonobjective" mind "acts" on matter? Most physicists say this is impossible. As I will show in the next chapter, recent theories and experiments suggest otherwise. And perhaps if the mind is partially quantum, we may someday come to understand how that helps it invent topology.

13

THE
QUANTUM BRAIN?

I am hardly the first person to assert that consciousness may be related to quantum phenomena. In 1989, the physicist Roger Penrose, in *The Emperor's New Mind*, proposed that consciousness is related to quantum gravity, the still missing union of general relativity and quantum mechanics. Here I will take a different tack and suggest that consciousness is associated with a poised state between quantum "coherent" behavior and what is called "decoherence" of quantum possibilities to "classical" actual events. I will propose that this is how the immaterial—not objectively real—mind has consequences for the actual classical physical world. I warn you that this hypothesis is highly controversial—the most scientifically improbable thing I say in this book. Yet as we will see, there appear to be grounds to investigate it seriously.

Virtually all neurobiologists think conscious experience is associated with the non-quantum mechanical, fully classical behavior of interconnected sets of neurons passing electrochemical "action potentials" along the long cell structures called axons to stimulate action potentials in neurons downstream through connections called synapses. The hypothesis that quantum mechanics plays any role in conscious experience faces major challenges, including fitting into the manifold work of neuroscientists and their "classical" neurons.

Yet if it should turn out that that quantum mechanics is deeply involved in conscious experience, it might help resolve four longstanding philosophical problems. First is the problem of free will. Briefly, the concern here is "causal closure." If every event, mental or physical, has sufficient antecedent causes, then as Aristotle said, there can be no "unmoved mover." But free will is supposed to be just such an unmoved mover, free to do what it chooses, hence an "uncaused mental cause" of our actions. This led the seventeenth-century philosopher Spinoza, and others since him, to conclude that free will is an illusion. Nor is this problem addressed by current neurobiological theory, in which conscious experiences are functions of the classical, fully causal behavior of neurons or neural circuits. If the mind-brain behavior is fully causal, how can we have free will, an unmoved mover?

The second problem also concerns free will. We want our free will to choose such that we can be morally responsible for our acts. If consciousness is deterministic, then it seems we are not morally responsible. If it behaves probabilistically, then again it seems it cannot be morally responsible. This is a very difficult problem. But the quantum consciousness hypothesis suggests a conceivable first wedge that might help resolve it, for a partially quantum conscious mind might be neither deterministic, nor probabilistic. It just conceivably might be partially beyond natural law.

The third problem is the problem of mental causation. If we assume, with almost all neurobiologists, that consciousness is identical with classical neural behavior (the so called mind-brain identity theory), and *if that neural behavior itself is causally sufficient for subsequent neural behavior,* how can "mind stuff," mental experiences, cause subsequent physical events? Such mental causation seems to require some new kind of "spooky" causation.

The fourth problem is the problem of epiphenomena. If brain states are correlated with mental states or identical to them, and the fully classical, fully causal activity of neurons is causally sufficient to cause subsequent action of neurons, are mental experiences mere epiphenomena, "pushed around" by the causal activity of neurons but having no causal relevance of their own?

If consciousness is partially quantum mechanical, all four problems can just possibly be resolved. For example, the problem of causal closure, which seems to render free will an illusion, disappears because in an *entirely acausal* quantum mechanical account of conscious experience and free will, *there is no unmoved mover, for the quantum behavior is acausal.* The issue of causal closure therefore does not arise. As I will show, the other three philosophical problems, moral responsibility, mental causation, and epiphenomenalism, also seem to disappear or may be open to resolution.

Even if these issues are resolved, we still face the central difficult problem: awareness itself, the blueness of blue as we experience it, what philosophers of mind call qualia. Are we conscious? Is there a philosophical problem of other minds? Since I cannot share your qualia, how do I know you have them? On the other hand, how do I know you do not have them? Experiences are private, "first person" phenomena. I will not argue for consciousness in other minds; it would take too long. But even if only I were conscious, the problem of my qualia would still exist, as would the problem of how my mind acts on matter and whether it is an epiphenomenon.

Descartes began modern philosophy of mind. As he famously said, he could doubt everything except that he was doubting: "I think, therefore I am." Descartes was the first "dualist." He proposed that there were two kinds of "stuff" in the universe, res cogitans, experiences, and res extensa, material stuff. Virtually all philosophers of mind and neuroscientists today reject dualism on a variety of grounds.

We will have to proceed without reviewing four hundred years of familiar debate about consciousness. I will, however, briefly describe the mind-brain identity theory, both because it is currently the most acceptable philosophic view of consciousness and because it is the basis of most neurobiological research on consciousness. It is also the basis of my own more problematic coherent/decoherent quantum mind hypothesis. The mind-brain identity theory asserts that mental states, qualia, are literally identical to specific neural states. Thus the theory rejects both Descartes' mind-body dualism and epiphenomenalism. Mind and brain are real and are one thing. While I will adopt the mind-brain identity theory for this discussion, it has very deep problems, as we've noted, with respect to free will and mental causation. In addition, how the "meat" of neurons can be *identical* to experiences is, if you will allow me, mind-boggling. But then, every alternative to the mind-brain identity theory has deep problems as well. For example, in Cartesian dualism, we do not have any idea how mind acts on matter. In the extreme idealism of Bishop George Berkeley, mind and ideas are real, but matter is "sustained" as the experiences in the mind of God. (Samuel Johnson felt he could refute this idea by kicking a stone.) Berkeley is thus the opposite of St. Augustine, for whom matter is real and for whom we are conscious by direct connection to the mind of God. The philosopher John Searle was right when he said that not only do we not understand what consciousness "is," we do not even have an idea of what it would be like to understand what consciousness is. While I hope to make some progress with my improbable hypothesis about quantum consciousness, I will make no progress at all on the more fundamental issue of qualia.

Long ago, as a student of philosophy of mind at Oxford, I read many fashionable texts arguing that conscious experience did not exist but was a "ghost in the machine," to quote the title of a book by philosopher Gilbert Ryle. Our actions, said Ryle, are "dispositions," mere regularities between stimuli and behaviors. This was logical behaviorism, the philosophic cousin of psychological behaviorism, which sought to eliminate consciousness in favor of regular relations between stimuli and behaviors. In the end, the behaviorists hoped to make consciousness, experiences, and all other internal mental states disappear. But in stubborn fact, we are conscious beings and we do have internal mental states, even if we have no idea how these states arise.

Some philosophers of mind, such as Patricia S. Churchland, contend that consciousness is not a unified thing. Rather, as the philosopher Owen Flanagan notes, it includes sensations, perceptions, moods, emotions, propositional-attitude states, procedural knowledge, and large narrative structures. Yet Flanagan provides a host of reasons to think that we may still have a unified theory of consciousness across all these domains.

THE STRONG
ARTIFICIAL INTELLIGENCE
THEORY OF CONSCIOUSNESS

The strong artificial intelligence position holds that a sufficiently complex web of computing elements will spontaneously become conscious. When applied to the brain as a computing system, this argument has considerable appeal. Connect enough neurons and at some threshold of complexity, consciousness arises more or less automatically. This position may be true and eventually, perhaps, even open to empirical testing. Indeed, with respect to neurons as the computing elements, this view is close to the neurobiologist's view. But the strong AI position does not depend on the physical makeup of the network. Neurons or Tinkertoys, if they are capable of computing, should equally become conscious above some threshold of complexity.

I'm doubtful. Recall Philip Anderson's multiple-platforms argument, that computation is independent of the physical apparatus that carries it out. What if we used water buckets, filled or empty to represent 1 or 0 states? Consider a system of millions of buckets pouring water into one another. Is this sufficiently complex that the system would become conscious and have experiences? I don't think so.

The water-bucket system might even pass the famous Turing test, according to which, if the computer were hidden behind a wall, a normal human observer could not distinguish its output from that of a human. I cannot prove that the bucket system would not have experiences, but neither can I believe it.

The Turing test itself is a suspect measure or criterion of consciousness. In the Chinese room argument, Searle asks us to imagine him in a room with a narrow open window. He has a very complete dictionary of English–Chinese symbols. An "observer" passes Searle complex English sentences. Searle uses the dictionary and writes down the corresponding Chinese symbols, and hands the results back to the observer. It appears to the observer that Searle understands Chinese and can translate from English to Chinese. But in fact Searle understands no Chinese at all. He is merely mechanically using the dictionary. While I like Searle's argument, a colleague pointed out that, were we to ask Searle to translate "time flies like an arrow, but fruit flies like a banana" into Chinese, how would Searle use his dictionary to translate this little beauty of a counterexample? So Searle's Chinese room argument is some kind of clue, but not faultless.

The strong AI argument for a threshold level of computational complexity above which consciousness arises is commonly accepted by cognitive scientists and others. Despite my doubts, it remains logically possible. *But the strong AI position requires that the mind be algorithmic.* We saw in the last chapter that the mind is probably not algorithmic, at least with respect to categorization and the general framing problem, let alone the attempt to formulate quantum gravity. If this is correct, then strong artificial intelligence becomes deeply problematic. If the mind is often not algorithmic, and if the mind makes meaning while computational systems are purely syntactic, why should we believe that this restricted computational capacity, which appears to be surpassed by the mind, is what elicits the mystery of consciousness?

CLASSICAL NEUROBIOLOGY

In the last chapter we discussed the idea, held by some neurobiologists, that the mind need not be algorithmic but could remain a classical physical system. That is, just as planets orbiting the sun are not carrying out an algorithm, mind might be the classical causal behavior of the brain, with which it is identical, without being restricted to being algorithmic all the time. Such a view treats the brain as a complex classical dynamical system, whose noisy dynamical behaviors underlie and are directly correlated with the variety of conscious experiences. Cowan's mapping of predicted neural activities to experienced hallucinations is plausibly consistent with conscious experience as caused by, or "identical with," physically classical neural behavior—where "identical with" is the mind-brain identity theory.

The neurobiologists' research plan (to summarize very broadly) is to find neurons or subcircuits of interconnected neurons whose rate and manner of firing, when above some threshold, correlate with conscious experiences. Christof Koch calls this the neural correlates of consciousness (NCC). Any theory of consciousness will have to explain why some but, critically, not all neural firings are NCC. Next, it is Koch's view that conscious experiences are emergent phenomena, arising from such NCC. A key feature of current work is that the firing of some single neurons seems to be associated with specific aspects of an experience, such as the redness of a chair. If true, then NCC can include the behavior of single cells. If different cells or subcircuits in different regions of the brain give rise to experiences of different aspects of experience, then neurobiologists face what is called the binding problem. The binding problem asks how diverse perceptions occurring in distant parts of the brain are bound together into a single conscious experience. Among the possibilities are that the neurons are somehow linked into subcircuits; that consciousness itself accomplishes the binding by focusing attention on the diverse features of experience; and that a fast, broadly based neural rhythmic activity in the brain may bind various perceptions into a coherent conscious experience.

There is debate among neurobiologists about what the neural "code" is. Is it the mean firing rate of neurons that is associated with experiences, or more subtle features of that firing, or even the chemical behavior of neurotransmitters released at the synapses connecting neurons? The popular view is that the code lies in the mean neural firing rate.

The neurobiologists' results must largely be accepted by any theory of consciousness. The search for the neural correlates of consciousness is clear, clean science, the current state of the art. But the view that conscious experiences are an emergent property of the firing of single neurons or subcircuits of neurons does not tell us what consciousness "is," nor how it "emerges." Further, classical neurobiology directly faces the familiar philosophical problems of free will, mental causation, and epiphenomenalism.

CONSCIOUSNESS AND
QUANTUM MECHANICS

Penrose and I both suspect that consciousness depends on some very special physical system. Unlike the strong artificial intelligence hypothesis, which is agnostic to the physical basis of the computation that attains consciousness, we suspect that consciousness arises only in very specific physical systems. Many neurobiologists share this view: they think the phenomenon is limited to physically classical neurons. With Penrose, I think it may instead be partially quantum mechanical.

The idea that the human mind is nonalgorithmic raises the *possibility* that it *might be acausal*, rather than a causal "machine," and the only acausal theory we have is quantum mechanics. Therefore, the mind may be partially quantum mechanical. This is a hypothesis rather than a deduction. Being nonalgorithmic is not the same as being acausal. As noted, a planet in orbit around the sun is not performing an algorithm, but it is nevertheless a classical physical system. Thus the apparent nonalgorithmic character of the human mind does not imply that the mind *must* be quantum mechanical. It may still be classically causal, just not algorithmic. If so, we will have to have a very serious discussion about what kind of classical system can do what the human mind seems to do. Remember our incapacity to finitely prestate all possible functionalities of LegoWorld objects, or screwdrivers (or of any other object)? Yet we constantly find new functionalities. How? How do we "play" and understand different nonreducible Wittgensteinian language games, such as legal language and ordinary human action language? Of course, a partially quantum theory of consciousness and mind will have to answer the same question.

When I suggest that consciousness is partially quantum mechanical, the idea that drives me concerns the *transition* from the quantum world of merely persistent *possibilities* to the classical world of *actual physical events*. As I describe more fully in a moment, in quantum mechanics, the Schrödinger wave equation is a deterministic equation that propagates in space and time. In the Copenhagen interpretation of quantum mechanics and the Born probability rule, the amplitude, or height, of the wave is thought of as the amplitude of a "possibility," so the Schrödinger wave describes the propagation of mere possibilities. It turns out by the Born rule that the square of that amplitude can be interpreted, by the Copenhagen interpretation of quantum mechanics, as the *probability* of a quantum process measured by a classical measuring apparatus, such as a specific polarization of a photon. In the propagation of the Schrödinger wave alone, no *actual events* occur. It is only possibilities that propagate and have amplitudes. And so quantum mechanics is entirely *acausal* on this interpretation. There is no "cause" for a specific actual measured event of a radioactive decay, it simply had an acausal probability of being measured to occur when and where it occurred.

But how does the quantum system of pure possibilities give rise to actual events, actual classical behavior? According to the Copenhagen interpretation when a physicist measures an event, say the position of a photon, using a *classical* measuring device, the "wave function" of possibilities "collapses" to an actual event, whose probability, again, is given by the square of the amplitude. Thus the Copenhagen interpretation already divides the world into quantum and classical realms. The classical world, acting on the quantum world via measurements, causes the quantum possibilities to collapse to some specific classical event. Of course, this doesn't say where the classical world comes from in the first place, if the universe at root is fully quantum. For example, in the very early universe, everything was presumably purely quantum. Where did the classical world come from? Many are concerned about this problem, for which "quantum phase decoherence," discussed below, is the current favorite answer.

The Copenhagen interpretation of quantum mechanics is not the only one. The "many worlds" interpretation proposes that the universe splits at each quantum measurement into two parallel universes, and the Bohm interpretation proposes that quantum behavior is fully deterministic, but unknowable. On the Bohm interpretation, quantum behavior is not acausal. Few scientists accept these alternatives. The probabilistic interpretation, which remains the standard, is *acausal*. We are able to compute the probability of the outcome of a classical measurement, but there is no underlying cause for that event. What happens, just happens to happen, with no cause.

Some years ago, Murray Gell-Mann, my colleague at the Santa Fe Institute, asked me if I knew any quantum mechanics. "No," I answered. "You really should," he responded. I decided to learn something about quantum mechanics. It is worth pausing here to describe an example used by Gell-Mann's late colleague at Cal Tech, Richard Feynman, to convey the central mystery of quantum mechanics. It concerns the famous two-slit experiment and quantum interference, a concept we will need as we explore consciousness in more detail.

Think of a wall with two small slits in it. On one side of the wall place a flashlight, or "photon gun," which shoots photons towards the wall. Behind the wall is a photo-detector surface—say a film or fancier system. Consider first the case of the wall with one slit covered up, leaving a single slit open. We find just what we would expect: the photons "pass" through the open slit and hit the photo-detector surface, creating a "pile" of photon reception events rather like a mound of sand, brightest at the middle. It is important to know here photons of the same wavelength have the same energy and are identical. As a result, they make the same size spot.

If the first slit is now covered and the second slit opened, and the photon gun is used, we once again see a mound of photon reception events behind the second slit, again like a pile of sand, brightest at the middle and fading toward the edges.

What if both slits are open? What happens is fantastic. As we continue to fire the photon gun, say one photon per hour, light and dark curved lines appear, on the photo-detector, rather like the intersecting waves created by simultaneously dropping two pebbles in a calm pond. This intersecting wave pattern of light and dark regions is *quantum interference*. There is no classical behavior that can explain these findings.

The Schrödinger equation clearly describes this odd behavior. The equation is a linear wave equation. The waves emanate from the photon gun and yield waves of "amplitude." Consider first parallel water waves, which might approach the two slits and, on passing through, create two spreading semicircular waves traveling toward a beach beyond the slits. These two water waves would interfere, with two peaks summing to a higher wave at some points, and peaks and troughs canceling each other at other points, creating an interference pattern when they arrive at the beach. Similarly, the Schrödinger wave amplitude through the two slits creates two spread-out wave patterns that interfere and then hit the photodetector, creating the interference pattern. As Feynman points out, the response of the photodetector is identical whenever a single photon hits it. So the interference pattern can be explained by thinking of the photon as both a wave and a particle passing simultaneously through both slits. The wavelike nature of the photons passing simultaneously through both slits is what creates the interference pattern. Feynman invented a marvelous equivalent "sum over histories" theory in which the photon, considered as a particle, simultaneously takes all possible paths to the photodetector. These paths interfere quantum mechanically constructively and destructively with one another to create the observed interference pattern.

Now the same idea, expressed in more mathematical prose: a sequence of water waves has a frequency, a distance between the peaks divided by the wave velocity. Thus we can conceive of a "phase" of the cyclic behavior of the wave, which varies periodically as the wave passes by. In the Schrödinger equation, a similar phase is associated with the equation's amplitude wave. Just as water waves interact to create higher waves or cancel each other, so, too, for the Schrödinger equation. The positive summing of crests and canceling of crests and bottoms in the two slit experiment is what creates the constructive and destructive interference pattern on the photodetector. This summing of crests and bottoms and, in general, the entire wave pattern with its phase information is the centerpiece of quantum behavior. It will be critical to our discussion of passing from quantum to classical behavior that quantum interference requires that *all the phase information* at each point on the photodetector be available so the wave peaks and troughs can add together properly to create the interference pattern.

The transition from quantum to classical behavior has been explored in great detail with abundant debate. Currently the theory of "decoherence" is the favorite candidate to explain passage from the quantum world of possibilities to the classical world of actual events. Decoherence is based on loss of phase information.

The decoherence theory is some twenty years old. It says that as a quantum system interacts with a quantum environment, such as a "bath" of quantum oscillators, that system's phase information becomes scrambled, in part due to becoming "quantum entangled," as explained briefly below, with the environment, *is lost*, and cannot be reassembled. With the loss of this phase information, quantum interference in the system cannot occur and classicity—some actual physical event—emerges from the "fog" of propagating possibilities. This interaction of a quantum system with its environment acts somewhat like the classical measuring device in the

Copenhagen interpretation: in some sense, partial measurement of the quantum system by its quantum environment leads to a partial loss of phase information, hence the loss of quantum coherence and the possibility of quantum interference, and thus the gradual onset of decoherence. "Gradual," in this case, can be on the order of a thousandth of a trillionth of a second, a femtosecond or very much longer, depending on how strongly the quantum system couples to the quantum environment. The existence of decoherence is well established experimentally and in fact is the bane of current efforts to build quantum computers.

I will make use of decoherence to classical behavior as the means by which a quantum coherent conscious mind of pure possibilities can have actual classical consequences in the physical world. This will be how mind has consequences for matter. Note that I do *not* say "how mind acts on matter," because I am proposing that the consequences in the classical world of the quantum mind are due to decoherence, which is *not itself causal* in any normal classical sense. Thus I will circumvent the worry about how the immaterial mind has causal effects on matter by asserting that the quantum coherent mind decoheres to have *consequences* for the classical world, but *does not act causally on the material world.* As we will see, this appears to circumvent the very old problem of mental causation and provide a possible, if still scientifically unlikely, solution to how the mind "acts" on matter.

It is important to stress that decoherence in quantum systems with many interacting quantum variables is still only partially understood. How decoherence actually happens is deeply important to our topic, and on the edge of current physics. It is known that if the quantum environment is made up of a "bath" of what are called quantum oscillators, the quantum analogue of many frictionless pendulums, then decoherence approaches classicity for all practical purposes (FAPP). But if the environment is what is called a quantum spin bath, then it has been shown that the entire system does not decohere to classicity, and some quantum coherence remains.

In sum, everything depends on how decoherence occurs, which depends upon the actual quantum physical system, its quantum or quantum plus classical environment and all its details. Decoherence to a quantum oscillator bath in a mathematical model of decoherence can assure classical behavior, FAPP, but need not assure classical behavior with a quantum spin bath. We are only beginning to understand how decoherence happens. Further, some theorists think that decoherence may not require the quantum system to interact with an environment at all. It may be intrinsic, and due to the relativistic curvature of spacetime or to events on the Planck length scale of 10^{-33} centimeters.

I will base my theory on the view of decoherence as due to interaction of a quantum system with a quantum environment—or a quantum plus classical environment—perhaps something like a quantum oscillator bath, the loss of phase information, and the emergence of classical behavior, FAPP.

The cornerstone of my theory is that the *conscious mind is a persistently poised quantum coherent-decoherent system, forever propagating quantum coherent behavior, yet forever also decohering to classical behavior.* I describe the requirements for this theory in more detail below. *Here, mind—consciousness, res cogitans—is identical with quantum coherent immaterial possibilities, or with partially coherent quantum behavior, yet via decoherence, the quantum coherent mind has consequences that approach classical behavior so very closely that mind can have consequences that create actual physical events by the emergence of classicity. Thus, res cogitans has consequences for res extensa! Immaterial mind has consequences for matter.*

More, in the poised quantum coherent–decoherent biomolecular system I will posit, quantum coherent, or partially coherent, possibilities themselves continue to propagate, but remain acausal. This will become mental processes begetting mental processes.

Some physicists will object to my use of the word *immaterial* with respect to quantum phenomenon. They would want to say instead, "not objectively real," like a rock. I am happy to accept this language, and will use *immaterial* to mean "not objectively real."

There are many possible alternative hypotheses about how quantum and classical worlds might interact in consciousness. For example, Penrose and Stuart Hameroff have suggested quantum phenomena in microtubules.

My own rather specific prototheory, now discussed in more detail, posits a nonmaterial quantum coherent mind that has consequences for the behavior of physical matter, res cogitans and res extensa united in one physical/biological theory. When an electron and a photon interact, they remain quantum. Indeed, this is the basis of the famous quantum electro-dynamic theory, known to be accurate to eleven decimal places. When two rocks collide they remain classical. Quantum interference does not arise if a stream of rocks is hurled at the two slit setup. Is it possible, then, to con-struct, or evolve, a system that remains poised between a largely quantum coherent state, where the Schrödinger wave propagates in what is called a "unitary" way, explained below, and a persistent decoherence?

Is it possible to have a real physical system that is able to remain poised in-definitely between parts that propagate quantum coherent linear Schrödinger wave behavior, and parts that decohere towards classical behavior where actual events happen in the world? Can such a poised system exist? Can it exist at any temperature, say, near absolute zero, and can it exist in a quantum system with many quantum variables or quantum degrees of freedom? Does such behavior require a quite dense and highly diverse set of quantum processes such as might be expected in something as complex as a cell in the brain? It is entirely open to research. The second question is whether such a system could remain poised between quantum coherence and decoherence at body temperature. Most physicists would claim immediately that this is not possible because no quan-tum coherence could persist at body temperature; decoherence would happen almost instantaneously. A system persistently poised between coherence and decoherence would thus be impossible.

But there is stunning recent evidence on this point, coming from studies of the molecules involved in photosynthesis, the process by which plants capture sunlight and turn it into chemical energy. The critical molecule for capturing photons is chlorophyll, and the protein that holds it is the antenna protein. The new evidence shows that when chlorophyll absorbs a photon of light energy and processes it with extraordinary efficiency ultimately to chemical energy, chlorophyll does, in fact, *maintain a quantum coherent state* for a very long time compared to chemical-bond-vibration frequencies. Quantum coherence can exist for about 750 femtoseconds, compared to the 1 to 1.5 femtosecond frequency of chemical-bond vibrations. This quantum coherent state mediates the very high efficiency with which plants and cer-tain bacteria convert light into chemical energy. More astonishingly, there is now direct evidence that the antenna *protein that holds the chlorophyll actually acts to sustain the quantum coherent state* against degradation due to decoherence. Otherwise concluded, the antenna protein suppresses decoherence, either by *preventing* it, or *reinducing coherence* in decohering parts of the chlorophyll molecule. Even more importantly, since the very high efficiency of transfer from light energy to chemical energy is critical to life, these results suggest very strongly that natural selection has acted on the antenna protein to improve its ability to sustain the quantum coherent state. Suddenly, the hypothesis that long-lived quantum coherent states may exist at body temperature is not at all ruled out.

The third question, of course, is what the proposed system in the brain might actually consist of. The fourth is how it could have evolved. None of the answers are known. But remember that chlorophyll and its antenna protein *did* evolve.

I now will list some of the requirements for and support for my poised quantum coherent/decoherent mind theory.

First, let's explore the acausal behavior of a quantum electron in a classical "box" under the Copenhagen interpretation of quantum mechanics. This box is a mathematical idealization used to solve the Schrödinger equation for the behavior of the electron in the box. The box provides what are called the mathematical boundary conditions, and confine the quantum electron. Here one pretends that one can isolate the quantum system—and it really is just a pretense. As Zurek points out, we cannot "isolate" the quantum system, because information "leaks" out of the box via quantum tunneling. But ignoring this issue for the moment, the canonical way physicists approach a problem like this is to use the mathematical boundary conditions consisting of the boundaries of the box and the amplitude of the wave at the boundaries. They then solve the Schrödinger equation, which has its own Hamiltonian function, for the amplitudes of the electron at various spatial positions inside the box, where the assumption that the electron does not escape the box guarantees that the solutions will have specific spatially definite distributions called eigenfunctions, each with its own energy level. The eigenfunctions are discretely different from one another. They square the amplitudes at each point in space, and calculate the probability that the electron, if it were to be measured with a classical apparatus, would appear at each point in the box. This probability is entirely acausal. There is no underlying cause that causes the electron to be there, only the probability that comes from squaring the Schrödinger amplitude. This acausal feature of the standard interpretation of quantum mechanics is central to my theory.

Second, if there is to be a persistent poised state between quantum coherence and decoherence, and the total system has a finite number of "degrees of freedom" (the physicist's fancy phrase for the ways that things can change), then it must be logically and physically possible for some degrees of freedom that are beginning to decohere to be "made" to recohere back to a quantum coherent state. Otherwise, if decoherence were irreversible, ultimately all the degrees of freedom would decohere and no further quantum coherence could persist.

A very recent theorem from quantum computing suggests that this reversion of partially decoherent "degrees of freedom" to fuller coherence is possible. Here is the outline of the idea. A quantum computer consists of quantum bits, or qubits, which might be atoms or ions in an ion trap. Now in the dynamics of the quantum computer, some of the qubits interact with the environment and begin to decohere. This may induce error in the quantum computation. The new theorem says that if there is redundancy in the computation, say five qubits where before there was only one, error correction can be applied to one of the five as it begins to decohere. It can be made coherent again. A measurement of the total quantum system is taken which measures the error itself, that is the decohering degree or degrees of freedom, but not the quantum computation. Thus this measurement will not collapse the wave function of the computation. But if *information* is injected into the system, this information can be used to cause the decohering qubit, or degree of freedom, to recohere. The error can be corrected. From our point of view, this implies that, in principle, quantum degrees of freedom can start to decohere, then be made to recohere. *If so, a poised state persisting between "largely" coherent and partially decoherent quantum variables looks possible.* In addition, as noted, quantum computer physicists speak of a protected subset of degrees of freedom that remain coherent. Such protected degrees of freedom may remain available to sustain the coherent state.

Quantum computer theorists study this quantum error correction by devising hypothetical quantum computer circuits that they could, in principle, couple to the quantum computer's qubits. When the decohering qubits are corrected by the outside circuit, the decoherence is transferred to the outside circuit. A tiny beginning of an approach to think about such circuits as realized in actual molecules has been suggested by physicist Philip Stamp. Thus, experimental realization of quantum error correction in a molecular system that is partially coherent may be possible.

In addition, there is a developing area of physics called quantum control. Here laser light pulses directed at a quantum system are able to alter the phases of the quantum system. So it is, in principle, possible, again by adding energy and information, here the laser pulses, to alter quantum phases and perhaps possible to "correct" decoherence.

Third, the idea that an injection of information can protect or return decohering quantum variables to coherence suggests how quantum behavior might persist at body temperature. Injection of information implies that the quantum system is open thermodyanamically to the transfer of matter and energy. As open thermodynamic systems into which matter and energy and information can flow, cells may have evolved the ability to maintain coherent or near coherent behavior. As one physicist told me, it no longer appears that sustained quantum behavior at body temperature is excluded in principle.

Fourth, the long-lived coherence recently established in chlorophyll molecules and the antenna protein shows that such coherence is possible, at least for 750 femtoseconds at 77K. It is almost certainly true that this coherence is related to chlorophyll's efficiency. The superposition of the linear solutions of the Schrödinger wave equation permits the chlorophyll system to *simultaneously explore in parallel all possible pathways* from the initial excited state to the low energy state of the "reaction center" where the photon's energy is used to synthesize ATP. Thus the parallel simultaneous quantum coherent search for this low energy state can be far more efficient than a serial, physically classical, one-at-a-time search through the intermediate possible steps downward from the first excited state to the activation center state. It is this parallel, simultaneous quantum search over the energy landscape that is thought to give the amazing 95 percent efficiency. If we could learn to do this technologically, we might solve our planet's energy problems.

Still, this long-lived coherence lasts only 750 femtoseconds, where a femtosecond is about a trillionth of a second. Neural events such as action potentials are a million times slower. If quantum coherence is to be part of an account of consciousness, it would seem to require mechanisms that suppress decoherence, restore coherence, or, by weak coupling or other means, allow coherence on time scales of milliseconds or more. A molecular system capable of remaining poised on long time scales between quantum coherence and decoherence, with recoherence of decohering degrees of freedom, could, in principle, achieve such long time scales of quantum coherence. While no such system is known at present, no physical law precludes its existence. Finding such a system would be a triumph. I will describe one potential cellular system that just might be able to sustain a quantum coherent and partially coherent state, open to theory and experiments, below. Neurons may already do it.

Fifth, we always use our most complex systems as our models of mind. Currently that system is quantum computation. The two core ideas of quantum computing are as follows: (1) While a classical computer can calculate only one solution at a time, a quantum computer relies on the superposition of states that arises from the Schrödinger equation. The quantum computer can process all these superpositions at once, hence *simultaneously "computing" all the solutions available to it.* If there are N qubits, each with only 2 quantum states, the system can compute 2^N solutions at the same time. (2) The quantum algorithm uses the constructive and destructive interference of the quantum wave, the adding of peaks and canceling of peaks, *to increase the amplitude in the vicinity of the correct solution.* That is, the quantum computer, given the algorithm, not only computes all possible solutions simultaneously and in parallel, but it magnifies the probability that the proper solution will be found. This is stunning. Perhaps something similar allows quantum mind to focus on desired solutions in the classical world.

These features of quantum computation raise the possibility that a poised quantum-classical mind system could process the entire set of sums and differences of the wave equation, and tune the interference to increase the probabilities of good eventual classical behavior. The mind thus searches a vast space of possibilities to create a "good" physical response.

Sixth, decoherence is a real phenomenon and understanding it is essential to my theory, but such understanding is at the frontiers of physics. How decoherence actually occurs in different concrete physical or biological systems is only slightly understood.

At present physicists studying decoherence use differerent sets of equations such as the quantum oscillator bath model, which in the decoherent limit becomes Newton's equations, for how amplitudes and probabilities decay to classical behavior. But again, decoherence depends upon the physical details, for a quantum system coupled to a quantum *spin bath* partially decoheres, but not to classical behavior. In addition to these methods, physicists use what are called Markov models of decoherence.

Interestingly, the physicists and quantum chemists sometimes "cheat" and use a quantum coherent model for a small part of the system and a classical model for the rest of the system, and mathematical rubber bands to connect the two parts. In such models, common in quantum chemistry, *no account at all is taken of decoherence.* This itself seems to imply that quantum chemistry, in which decoherence is virtually certain to occur, must be modified to include decoherence. This might become an entire new branch of quantum chemistry and more generally of quantum mechanics, where decoherence in multiparticle quantum systems is poorly integrated into theory and experiment. The inclusion of decoherence may only make minor modifications to current predictions, or perhaps radically new predictions.

The chlorophyll-antenna protein system is our best current model of such a system. If we can develop an adequate theory of coherent and decoherence behavior in complex molecular systems, we can study how the antenna protein suppresses decoherence and also begin to explore other issues. Ultimately, we may be able to predict the structures of organic molecules, including protein and lipid systems, able to maintain long-lived quantum coherent states; study evolved proteins, lipids, and neural membrane or other macromolecular assemblies in cells to see if they harbor quantum coherent behavior; design quantum coherent organic molecules and macromolecules and study macromolecular systems in cells at body temperature; and even map out my hoped-for poised state between quantum coherence and decoherence in an open thermodynamic nonequilibrium cell system.

Seventh, well-known facts about cells and recent quantum chemical theory raise the possibility that *vast webs of quantum coherent or partially coherent degrees of freedom may span large volumes of a cell.* This is one conceivable model of my hoped-for poised state between quantum coherence and partial decoherence in an open thermodynamic nonequilibrium system. What I now describe is partially known, and partially my own scientific speculation. Therefore, I warn the reader to take caution. What I shall describe, however, constitutes achievable theory and is, in principle, testable. First it is critical to realize that a cell is a highly crowded, densely packed, and organized arrangement of myriad macromolecules such as proteins, other macromolecules, water, ions, and small molecules. The water in the close vicinity of these macromolecules is *ordered* for some distance. Examples are now well known in which a single protein with a cavity can contain one or a few water molecules held in place, or ordered, due to what are called hydrogen bonds between the water molecules themselves bonding with the protein. These may support more quantum coherence than we yet know. For example, quantum chemical computations show that quantum coherent electron transport through *two different ordered water molecules* from an electron donor to an electron acceptor part of the protein can occur. Because two pathways exist, both can be taken simultaneously, as in the two slit experiment, and quantum coherent behavior occurs. Furthermore, quantum coherent electron transfer within protein molecules is well established, at least in quantum chemical computations. It is also known by quantum chemical calculations that if one examines two different proteins at different distances apart, and studies the electrical conductivity from one modeled protein to another as a function of distance, that something wonderful occurs. An ångström is 10 to the negative 10 meters, and the proper scale for atoms and molecules. If the two quantum-chemistry-modeled proteins are separated by between 12 and 16 ångströms, and the computational model uses that space to place ordered water molecules between the two proteins, quantum coherent electron transfer occurs, and electrical conductivity between the proteins remains high and almost constant as the distance between the two proteins increases from 12 to 16 ångströms. Beyond about 16 ångström separation, so many water molecules can enter between the two proteins that they are no longer ordered by hydrogen bonds and behave like liquid water, and electrical conductivity falls off rapidly and exponentially with increasing distance between the proteins beyond 16 ångströms. All this is consistent with the view that quantum coherent electron transfers occur at about 12 to 16 ångströms between the proteins via ordered sets of one or more water molecules between the proteins.

So much is reasonably well established. Now I need to describe percolation theory, which is also well established. Consider a white-tiled floor, with say, square tiles. Now randomly replace a few white tiles with a few black tiles, so the fraction of black tiles is 1 percent. Ask whether you can walk across the tiled floor, tile by tile, stepping only on black tiles. Of course you cannot. Now if all the tiles were black, you could walk across the floor tile by tile stepping only on black tiles. Let the fraction of black tiles randomly placed on the floor increase from 1 percent to 50 percent to 59 percent. Magically, at the critical "percolation" threshold of 59 percent for a square tile array, you *can* walk across the floor tile by tile stepping only on black tiles, along at least one pathway across the floor. Define the size of the largest connected black tile "cluster." It turns out that below the critical percolation threshold of 59 percent black tiles, as the size of the floor and number of tiles increases, the size of the largest connected black cluster does *not* increase in proportion to the size of the floor. But at 59 percent and above 59 percent black tiles, the size of the largest black cluster *does* increase in size with the area of the floor. Thus, for a large floor a "vast" network of black tiles *spans*, or *percolates*, across the floor, leaving white tiles as well.

We now pass to my own scientific speculations, which are completely open to theoretical and perhaps experimental test. The cell is densely crowded with proteins, other macromolecules, ions, and ordered and disordered water. Suppose the packing is such that the typical distance between proteins and other macromolecules is 12 to 16 ångströms. This is open to direct test, for example using what are called freeze-fractured cells, probed with an atomic force-field microscope. Now, proteins are very large compared to water molecules. Imagine very many ordered water molecules between any two proteins or other macromolecules, employ current quantum chemistry, and calculate the *number* of possible *quantum coherent electron transport pathways* between different points on the two proteins via two nearby ordered water molecules. Let's call each such quantum coherent electron transport pathway a "bridge" between the proteins, and color it red in our mind's eye. We don't know the number of red bridges per pair of proteins yet, but we can know, at least by quantum chemical calculations. Calculate this for many pairs of proteins to get the distribution of the number of such red bridges between different pairs of proteins. Given that, we can calculate from percolation theory whether the *red bridges form a percolating vast connected web between the macromolecules in the cell.*

I note that electron transport is "directional" from a donor to an acceptor site, depending upon what is called the redox potential. A donor-acceptor pair has the property that the electron moves from the donor to the acceptor. However, local changes in the environment, for example by movement of protein side chains, can alter the electronic "cloud" charge distribution near donor and acceptor pairs, changing the donor into the acceptor and vice versa. When this happens, the electron will tend to move in the opposite direction. This implies that our red bridges should have arrows on them, showing the direction of electron flow, but the directions of the arrows can change as the local electron cloud environment changes. Due to such fluctuations, in principle, cyclic electron transport pathways can occur.

Next, let's study the probability that quantum coherent electron transport across the proteins themselves can *link the two ends of two red bridges ending on the same protein.* It is well known that quantum coherent electron transport within a protein can occur. If quantum coherent pathways connect two red bridges, color such intraprotein pathways red as well. Now we can ask if a vast red linked quantum coherent electron transport web, between proteins and through the proteins, spans large volumes of the cell. If so, we would expect that a single wave function for that web would exist and that quantum coherent behavior would occur with respect to electron behavior in the red web.

The next step notes that the ordered water molecules can move, often on a time scale of several billionths of a second. In the analogy with the white and black floor tiles, imagine removing black tiles from the floor at some rate. Let them be immediately replaced with white tiles. Then with some billionths of a second time scale, the time scale for such motion of ordered water, replace the recently placed white tiles with the free black tiles. This, in our chemical context, is the movement of one ordered water molecule from its ordered position, and its replacement a moment later by another water molecule. Now calculate, given the replacement rate, and lowered number of red bridges at any instant, whether we still get a red percolating quantum coherent electron transport web in the cell. Presumably, since the ordered water molecules move, as do the proteins and other macromolecules, and small molecules not yet considered but through which quantum coherent pathways can exist, the coherent web changes over time as new water molecules become ordered. But it may always be the case that a percolating vast quantum coherent web is present.

At this point, the above sketch of a theory is feasible to compute. Finally, add decoherence and perhaps potential recoherence, for example by a molecular realization of quantum error correction or other means, possibly related to quantum control, in the open thermodynamically non-equilibrium cell. We only approximately know how to do this for a complex system of proteins, other macromolecules, water, and ions as they move about in the cell, and the electron clouds around thousands of electrons move and redistribute. At the end of such a theoretical effort, we can hope to test theoretically whether, in the presence of moving molecules and decoherence, without or with a molecular realization of quantum error correction or quantum control from within the cell, we would expect a quantum coherent web spanning some or all of the cell, none of the cell, or small clusters of quantum coherent electron behavior among some proteins and macromolecules and water bridges plus "through space" quantum transfer. In addition, given a decent theory of decoherence, and hopefully, means of recoherence, we can test whether a connected, long lived, partially coherent, partially decoherent web spans the cell and remains poised between coherence and decoherence, and whether the poised web forms and reforms, but one is always present, with partially coherent connections maintained over time as water molecules and proteins and other macromolecules and ions move.

I note that there may be nothing special about electron transport. We could consider other quantum variables such as electron spin, phonon, and other quantum degrees of freedom that link strongly or weakly to the other degrees of freedom, and thus may decohere rapidly or slowly, and ask the same questions about percolating webs of quantum coherence in cells.

A potentially important issue arises: *what is the physics of a persistently partially quantum coherent system?* Physicists have begun to study decoherence itself, but not yet possible long lived partially coherent, partially decoherent quantum systems. So we may have much to learn about the physics of such systems.

What should we make of these established biological facts about high molecular crowding in cells, possible grounds to expect quantum coherent electron transport within and between proteins, and my percolation theory approach to vast quantum coherent webs? Well, we don't know, but the theory seems, at present, worth doing. As I remarked, using freeze-fractured cells and atomic force-field microscopy, it might now be feasible to measure the distribution distances between adjacent macromolecules such as proteins in cells. If, in fact, the mean distance is on the order of 12 to 16 ångströms, this might suggest that natural selection has maximized quantum coherent intracellular web processes, just as it has presumably maximized the ability of the antenna protein to suppress decoherence.

Direct evidence for quantum coherence in electron transport may be possible. For example, measurements of electrical conductivity in cells might be one experimental avenue. If so, exposure to hypotonic solutions to swell the cells, like lettuce in pure water, would increase the mean distance between molecules and might destroy such coherent behavior. (*Hypotonic* roughly means that the solution has more water molecules per unit volume than does the cell. Hence, on exposure to such a hypotonic medium, the excess water diffuses into the cell and causes it to swell.) The quantum calculations for two model proteins as their distance increased found a "plateau" of roughly constant electrical conductivity as distance between the modeled proteins increased from 12 to 16 ångströms, then fell off exponentially. This plateau was attributed to quantum coherent electron transfer. One might hope to see a similar plateau as a function of increased average distance between intracellular macromolecules as cells swell in hypotonic media. Alternatively, Aurelien de la Lande has suggested molecular nanotechnology experiments that might approach this question. It might be hard to rule out classical explanations, however.

At a minimum, it seems fair to say that such sustained quantum coherent webs and sustained partially coherent webs do not seem physically impossible and deserve investigation. Such a web could, in principle, yield quantum coherent behavior on very long time scales compared to the mere 750 femtoseconds for chlorophyll. If we need to reach a millisecond or longer time scales to be neurologically meaningful, this could possibly do it.

There is yet another deep problem. How could quantum coherence in a small region, say, of a neuron, have spatial long-range effects? There is a further, perplexing feature of quantum mechanics that creates long-range correlations in real space, and possibly in real cells, including neurons. Pairs or larger numbers of quantum particles can become what is called "quantum entangled." After the particles are entangled, they can separate to arbitrary distances at speeds up to the speed of light and remain entangled, and if one of the particles "is measured" as having a given property, the other particle instantaneously has a corresponding property. This deeply puzzling feature of quantum mechanics gave Einstein the gravest concern, captured in a famous paper by himself, Boris Podolsky and Nathan Rosen in the 1930s. The implied instantaneous correspondence has now been confirmed experimentally, and is called nonlocality. Whether and how entanglement could give rise to long range quantum correlations in living cells or circuits of cells is essentially unknown.

More, how entanglement and decoherence work in spatial volumes with dense and diverse quantum processes such as might occur in cells is also still only poorly known. However, part of how quantum phase information spreads from a local quantum coherent system to the environment and rest of the universe is via entanglement, and part of decoherence of a local quantum system is due to loss of phase information of entangled quantum particles as they escape the local system and cannot be recovered. For example, electrons exchange photons. An electron in a quantum coherent web might emit a now entangled photon which flies off into the universe at the speed of light, with no way for the cell to recover it.

Suppose we should find quantum coherent and partially coherent behavior spanning within and perhaps even between virtually all nerve cells with their well studied synapses between neurons creating neural circuits, synaptic vesicles, and neurotransmitters that allow presynaptic neurons to stimulate or inhibit firing of post synaptic neurons. Then we would face the same problem that Christof Koch faces. Only some neurons seem to be the neural associates of consciousness; others are not. We would need to find why this might be true on this quantum coherent-decoherent model of mind.

Finally, I comment on the possible coupling of "classical" neural behavior to possible quantum coherent behavior in webs or otherwise. In an action potential, about 8,000 ions enter or leave the neuron through one membrane channel per millisecond. A femtosecond time scale, where rapid decoherence occurs, is a million times faster. So one ion enters or leaves the cell about every hundred million femtoseconds. It does not seem impossible for potential coherent or partially quantum coherent webs in cells to form and reform many times as ions enter and leave a cell. And, of course, the entry and exit of diverse ions, mostly sodium and potassium, would alter both the hoped for quantum coherent behavior, sustained partially coherent behavior, and decoherence processes.

All of the above discussion does not render my hypothesis at this point anything more than physically possible, potentially testable, and potentially interesting. Even if it should be true, the problems of the "neural code," the binding problem, and others will arise again in this new context. I again caution the reader that this hypothesis remains scientifically the most questionable idea in this book. But something like it might be true.

The Poised Quantum Mind Hypothesis Again and Philosophic Issues

Suppose consciousness is associated with a system persistently poised between quantum coherence and decoherence. If this coherence or partial coherence is associated with consciousness, we may explain how immaterial mind can have classical physical consequences.

In this sense, this idea is reductionistic. But as we have seen in previous chapters, physics could not have predicted the emergence of the heart in the evolution of this specific biosphere. The physicist cannot deduce the occurrence of specific Darwinian preadaptations, nor can the physicist explain—deduce—the evolutionary emergence of a molecular structure such as chlorophyll that captures sunlight to make sugar efficiently via quantum mechanical behavior. In the same way, the evolutionary emergence of a poised quantum coherent–decoherent system in a specific evolved multiparticle molecular system in the brain cannot be deduced or explained by physics alone. Such a poised system is ontologically emergent.

How would we begin to test whether a quantum coherent state (or for that matter a classical state) corresponds to conscious experience? Partly by seeking the neural correlates of conscious experience, as Koch and others are now doing. But even to say "corresponds" is to take a philosophical position. That position is, roughly, the mind-brain identity theory, in which first person experiences are identical with specific brain states. In the present discussion, the correspondence is between partially quantum coherent brain states and experience, and between related decoherence and classical physical events in the brain or body. We might begin to believe this theory were we to find ion channel or other proteins, membrane lipids, or both, or something else, that, like chlorophyll and its antenna proteins, or my hoped for percolating webs of quantum coherence or partial coherence in neural cells, that were capable of very long-lived, presumably spatially distributed quantum coherent states. We might then look for patterns of decoherence and recoherence affecting neurons downstream. If we found that these quantum coherent states and decoherence correlated with conscious experience in neuronal correlates of consciousness, such as single cells known to correlate with specific aspects of conscious experience as described above, and decoherence affecting the classical aspects of brain and body, this would constitute positive evidence. Such a research program is not unimaginable.

I remarked at the outset of this chapter that contemplating a quantum theory of consciousness is not foolish, not only because such a system might be more powerful than a strictly classical one, but because a quantum hypothesis might help address several deep philosophical problems that plague the classical stance. Like every other theory in the philosophy of mind, the mind-brain identity theory has serious problems—at least five, at least one of which currently seems insurmountable.

First, does mind act on matter? This is the question of mental causation: do my experienced intentions somehow cause my finger to move? In the standard classical mind brain identity view, the classical physical state of the, say, neural network *is identical* to conscious mind acting on matter. So we want to say the answer is yes. But some philosophers have argued that this is logically inconsistent. If we split the classical state into physical and mental aspects, and if the physical aspects are the sufficient causal determinants of the subsequent physical aspects, just how do the mental aspects "act" on the physical aspects? We have no idea. The way the mind acts on matter is both superfluous, as the physical aspect is causally sufficient already, and utterly mysterious. So, goes the argument, we have no theory for how the intentions of the mind act on the physical world. We seem blocked. Indeed, we seem forced to retreat to a position called epiphenomenalism, in which the "mind" is a side effect of the physical brain. The neural networks undergo their classical dynamics and events happen in the body. Worse, the physical events in the brain causally "push around" the mental experiences, which are mere consequences of brain activity and have no influence in the physical world. Our intentions have no effect on our physical or mental actions.

A second aspect of the mental causation issue is not how mind acts on matter but how mind acts on mind; how do thoughts beget thoughts on the mind-brain identity theory? Again, if we split the identical mind-brain states into mental and physical aspects, the physical aspects are causally sufficient for subsequent physical events. How then do the mental aspects cause subsequent mental aspects, such as the stream of consciousness? Again, mental effects on mental experiences, say in a stream of consciousness, seems to require spooky causes.

But the very statement of the "mental causation" issue is *purely classical, dealing with classical causes, and is based on causal closure of causally sufficient antecedent conditions.* Mind cannot act on matter, because brain already does. There is nothing left for mind to do, and no known means by which mind might do so in any case. Again, if brain states are already sufficient causal conditions for subsequent brain states, then there is nothing left over for mind to do in acting on matter. Even if there is nothing left for mind to do to matter, it would have to do it by unknown spooky means.

The argument above simply does not apply to a mind-brain identity theory based on consciousness as partially quantum coherent. The essence of the argument is that quantum coherent expression of the Schrödinger wave yields *possibility waves*, not *causes*. It is entirely *acausal* on the Copenhagen interpretation. Causal closure of the "physical" aspect of the mind-brain identity does not even arise, and the issue of "mental causation" becomes merely a confusing use of language. *The quantum coherent mind does not, in this view, "act" on matter causally at all. Rather, via decoherence, the quantum coherent state has consequences for the physical classical world.* Under this interpretation of the brain-mind identity, neither the "physical" brain with its quantum possibilities nor the "mind" *cause* the subsequent events in the physical world in anything like a Newtonian or Laplacian sense. Instead, decoherence to classicity occurs and has consequences for the physical world. Immaterial mind, in the sense of pure quantum possibilities, has consequences for the material world via decoherence.

If the above view is correct, we are not forced into epiphenomenalism by the arguments against mental causation with respect to the physical world. There simply is no such causation. The quantum brain does not push around experiences, because the acausal quantum aspects of the brain do not "push" at all. Thus it seems that a partially quantum coherent theory of consciousness, coupled with a mind-brain identity thesis, can allow mental experiences to have consequences on actual happenings in the physical world without invoking spooky mental causes of physical events. Mind has consequences for the material world by decoherence, a real phenomenon. There simply is no "pushing" in the poised coherent–decoherent quantum system. It is entirely acausal.

Finally, what of the second sense of mental causation, mental aspects causing further mental aspects, such as a train of thoughts? On a classical view of neurons, and the mind-brain identity theory, we seem stuck. If the classical mind-brain can be split into physical aspects and mental aspects, and the physical aspects are causally sufficient for subsequent physical aspects, we are left with no mechanisms by which mind can have consequences for mind.

But if consciousness is partially quantum coherent, then on a mind-brain identity theory, the entire split system of "physical aspects" and "mental aspects" of the identical mind-brain system are not causal at all. *Mental aspects unfold acausally into further mental aspects by the propagation of the quantum coherent state itself, perhaps with constructive interferences that focus quantum probability on desired solutions, as in quantum algorithms. No spooky mechanism is needed for such mental unfolding.* Alternatively, mental processes may give rise to mental processes by whatever processes occur in a partially coherent, partially decoherent system. This includes the possibility of quantum control of quantum phase information by *nonlinear quantum feedback* of the type known in Bose-Einstein condensation. Since all of these processes are not classical, again no spooky mechanism is needed for such mental unfolding.

Thus while the hypothesis of a quantum coherent mind is surely scientifically improbable at this point, it might solve a number of perplexing problems about mental causation and epiphenomenalism. It answers the question of how immaterial mind, again in the sense of mere quantum possibilities, has consequences for matter—not via classical causes but through decoherence. It answers how immaterial mind undergoes a flow of experiences.

The second and deepest philosophic problem is that of experience itself, qualia. This is sometimes called the "hard problem." How does all this fancy quantum coherent theory yield conscious experience? This theory makes no progress at all on the hard problem. Just as with the correspondence between classical neural behavior and experiences—Cowan's experienced hallucinations and patterns of neural activity, for example, or Koch's neural correlates of consciousness—the quantum-mind hypothesis tells us nothing about "how" quantum brain events give rise to first person experiences or what such experiences "are." And bearing in mind Donald Davidson's comment about anomalous causation in the mind, we may only ever find very simple cases open to clear correlations. It may be that this will be the best we can do. Were we to find, for example, abundant correlations of the quantum kind I hope for, we might still be left mystified about the emergence of first-person qualia in the living brain. Earlier, in considering the mind-brain identity theory with a classical view of the "meat" brain, I admitted that how "meat" and mental experiences could be *identical* was, to me, mind boggling. What if mind is not classical, not "meat," but persisting quantum possibilities in the quantum coherent brain? I am less boggled by the idea that experience and persisting quantum possibilities in the brain are identical. But I hasten to add that my own state of relative mind-boggledness is no criterion for scientific sense or philosophic sense. Nevertheless, somehow, I am less mystified because no one has any idea just what quantum possibilities "are," nor what experience "is." While I like it, this is still no answer, at least as yet, to the hard problem. Perhaps one day, it will become part of an answer.

The third problem is "free will." Here the problem is logical and based on causal closure. Causal closure asserts that everything that occurs has a sufficient antecedent condition or set of conditions. By "free will" we typically suppose an "unmoved mover," able to do anything it chooses. But this, goes the argument, violates causal closure. How does the unmoved mover manage to move? But this concern again seems to me to be completely classical. It is based on the idea that every event has preceding causes. But if mind is quantum mechanical and acausal, the concept of " unmoved mover" is moot. There is only the behavior of the quantum coherent/decoherent system with consequent classical events. In the quantum coherent behavior of the Schrödinger wave of possibilities there are no causes at all. There is no "mover." Thus there is no "unmoved mover" either. Causal closure does not apply. Hence the issue of free will as a prime mover in the classical causal sense does not arise. Thus it would seem that a theory in which consciousness is partially quantum coherent can embrace free will. As we saw earlier, it is the problem of the unmoved mover that moved Spinoza to conclude that free will is an illusion. A mind-brain identity theory, coupled with a quantum coherent theory of consciousness and decoherence with classical consequences, suggests that Spinoza may be wrong.

There is another aspect of the free will problem that a possible quantum mind-brain identity theory may, or may not, help resolve, but it should be stated. We want of a free will not only that it is free of causal closure, but that it is able to behave in such a way that we are responsible for our intended actions. On this version of the mind-brain identity theory, this requires that some appropriate systems of classical, quantum coherent, and decohering degrees of freedom of the brain can be identical to mind's intentions, which alter the way decoherence happens, hence what happens in the material world.

The philosophic problem is the following: If the mind is deterministic, then we do not have free will and are not morally responsible. If instead, our free will is merely probabilistic, then what our will chooses is merely chance, and again we are not responsible. Here and in the final note to this chapter, I broach the speculative possibility that for sufficiently dense and diverse multiparticle quantum process systems and environments, or mixed quantum and classical systems and environments and the universe, such as a brain cell, *the way phase information is lost to the environment may be unique in the history of the universe in each specific case. But then no compact description of the details of the decoherence process can be available, hence no natural law describes that detailed decoherence process. Moreover, the specific way decoherence happens in detail may matter for how free will chooses under intentions and comes to have consequences for the objectively real, that is, the classical, world.* The same would seem to be true of a persistent partially coherent mind.

In the case of such unique decoherence processes, the frequency interpretation of probabilities is moot. We cannot assign a probability. And on the Laplacian N door "ignorance" view of probabilities, we cannot assign a probability either, for we do not know N, the sample space. Rather, mind could be partially lawless, or contingent. This is at least a wedge in the philosophic problem of deterministic, so not responsible, or probabilistic, so not responsible. More, quantum processes in the brain could decohere to classical behavior and thereby modify the classical aspects of the brain-mind system. Thereby, this could alter the classical boundary conditions of the partially coherent quantum mind, altering its behavior and consequences for the classical world. Thus, we may confront a partially quantum analogue of the cycle of work in chapter 7, where cells do work to build boundary condition constraints on the release of energy that does more work, including constructing more boundary condition constraints on the release of energy in a Kantian propagating organization of process. Perhaps from some partially contingent, partially quantum elaboration of this hint, we can find free will that acts responsibly.

From the larger perspective of this book, consciousness is real; it is stunning; as far as we know, it is emergent in evolution; and we begin to have a few candidate accounts of how it may arise naturally, without a Creator God. Thus we have seen in this book the limitations of reductionism; plausible routes to the origin of life, agency, and meaning and value; and the persistent creativity of the biosphere, the economy, the human mind, and human historicity, and perhaps also consciousness. All, I claim, arose without a Creator God. If true, does this lessen our awe and wonder? Does it render these phenomena other than sacred?

14

LIVING INTO MYSTERY

For over three centuries, the reductionist model has resulted in many groundbreaking theories and successes in physics and beyond. By going beyond reductionism, I do not wish us to lose its continuing power for much of science. I, too, still dream of a final theory, with Weinberg. But that theory will not explain evolution, Darwinian preadaptations, economic evolution, cultural evolution or human history. In reductionistic physics there are only particles in motion, captured now in the standard theory and general relativity or perhaps string theory, should string theory succeed. But there are only happenings and facts. The reductionist confronts us with Weinberg's pointless universe in which the explanatory arrows always point downward to physics. Yet the emergence and evolution of life and of agency, hence value and meaning, cannot be reduced to physics. Although it may be too early to understand the true meaning of consciousness, there are reasons to believe that consciousness is fully natural. Moreover, if I happen to be right in my hypothesis that consciousness is a specific, evolved organization of quantum coherent and decoherent organic molecular systems in the brain that evolved organization and its behavior, like the evolved heart, will be ontologically emergent, and its evolved existence in the universe will not be reducible to physics.

Further, we have looked at the ceaseless co-constructing creativity of the biosphere evolving in part by Darwinian preadaptations, and their analogues in technological evolution. These preadaptations are radically unpredictable. And if we believe that a scientific law is a compact description beforehand of what will occur, then this evolution is not fully describable by "natural law." In the new scientific worldview I'm describing, we live in an emergent universe of ceaseless creativity in which life, agency, meaning, consciousness and ethics—discussed in a later chapter—have emerged. Our entire historical development as a species, our diverse cultures, and our embedded historicity, have been self-consistent, co-constructing, evolving, emergent, and unpredictable. Our histories, inventions, ideas, and actions are also parts of the creative universe.

There is in the co-constructing biosphere, economic evolution, and cultural evolution, a persistent individuation of subsystems and processes with diverse and ever modifying or entirely novel functionalities, that persistently fit with one another more or less seamlessly, and also fit into larger systems, hence come to exist and persist for periods of time, then evolve further. In organisms, the elaboration of organs and their processes that coevolved together to mesh their functionalities into evolving physiologies that sustain organismal life but evolve over time is roughly akin to species forming ecosystems where self-consistent community assembly occurs so that the species and abiotic environment form niches in which the same species can exist and sustain mutual life, yet evolve over time. In turn this is analogous to coevolving technologies of complements and substitutes that persistently fit together, such as the automobile, gasoline, motels, and suburbia, sustain economic activity, yet evolve over time. In turn this is analogous to the individuation of roles and rules in evolving societies and cultures that co-construct and coevolve their social roles and enabling and constraining rules in which our human cultures exist, persist, and change. It is within these cultures that our lives find much of their meaning. *None of this self-consistent co-construction seems to be what we mean as describable in its detailed becoming by natural law. Yet it is, in fact, truly happening all the time.*

Law "governed" and, partially beyond natural law, self consistently co-constructing nature itself has given rise to all that we have called upon a transcendental Creator God to author. Let God be our name for the creativity in the universe. Let us regard this universe, all of life and its evolution, and the evolution of human culture and the human mind with awe and wonder. It becomes our choice what we will hold sacred in this universe and its becomings.

T. S. Eliot once argued that Donne and other sixteenth-century metaphysical poets separated reason from the rest of our sensibilities. If this is true, then this fracture influences the deepest parts of our humanity. We have seen reasons why science may be limited in radical ways by the very creativity of the biosphere and human culture. If we only partially understand our surroundings, if we often truly do not know what will happen, but must live and act anyway, then we must reexamine our full humanity and how we manage to persevere in the face of not knowing. Reexamining ourselves as evolved living beings in nature is thus both a cultural task, with implications for the roles of the arts and humanities, legal reasoning, business activities, and practical action, and part of reinventing the sacred—living with the creativity in the universe that we partially cocreate. Because we cannot know, but must live our lives anyway, we live forward into mystery. Our deep need is to better understand how we do so, and to learn from this deep feature of life how to live our lives well. Plato said we seek the Good, the True, and the Beautiful. Plato points us in the right direction.

Reintegration of reason with the rest of our full humanity takes me far beyond my domain of expertise. But I believe we must try to do so in the light of the new scientific worldview I have been discussing.

Aristotle set the stage for the subsequent twenty-five hundred years of thinking about scientific knowing. He argued that scientific knowing is deduction from a universal premise and a subsidiary premise to a conclusion: All men are mortal; Socrates is a man; therefore Socrates is a mortal. This became fully enshrined in Newton's laws and subsequent physics. The laws are the universals. The initial and boundary conditions are the subsidiary premises, and the predicted future dynamical evolution of the system is the deduced conclusion. Notice again that this model fails for the co-constructing, partially lawless evolution of the biosphere by Darwinian preadaptations, and fails for the similar co-constructing evolution of technology and the economy by similar preadaptations—the unexpected discovery of novel functions. The "conditions of becoming" in the biosphere lead me to wonder whether and where such partially lawless co-construction may exist in the abiotic universe. Again, if a scientific law is, as Murray Gell-Mann suggests, a compact description, available beforehand, of the regularities that will unfold, then evolution of the biosphere by preadaptations and the economy by similar preadaptations are not describable by "law." In fact, as Aristotle already argued with his four causes, scientific knowledge fails for practical human action, where, he thought, final, formal, efficient, and material causes were necessary. But even Aristotle did not discuss the implications of the persistent limitations of our knowledge for how we manage to act anyway. How inarticulately we speak of practical action. Yet we live it every day.

In this regard legal reasoning was held to be the highest form of reasoning in the Medieval period. With the success of Newton, it lost its status to that of scientific reasoning. Legal reasoning covers practical actions. It adjudicates the reasons and motives we have for our acts in concrete situations, and does so by laws codified and interpreted over the years by precedent, legislation, or both. Indeed, if we contemplate an evolved body of law such as English common law, started in 1215 by the Magna Carta, it is a magnificent collective human achievement.

A close friend of mine has a son and daughter-in-law, Nathanial and Sarah, who have a seven-month-old little boy named Quinn with the most remarkably round, blue eyes. Recently we spent the evening together at a New Year's Eve party. Sarah was much involved in the ongoing conversation and celebration. But she was also instantly alert for any sign of alarm from her little son. I watched them in the mother-child dance of enfoldings and explorations. Sarah's understanding of Quinn and his needs is not something we can quantify. It reflects all of this young mother's humanity and her understanding of her son as a human being, an acting, sleeping, laughing, crying infant. But Sarah's humanity reflects 5 million years of hominid evolution, longer mammalian evolution, and probably further back than that. It's well known that if that mother-child bond is compromised in the first months of life, profound attachment issues often arise for the child and can equally affect the child's future well-being. Presumably we evolved in clans in which babies were held by someone, mother, aunt, sister, brother, father, for the first six or more months of life most of the time. If true, we should know it, for our methods of childrearing in the first world now disregard what may be our psychological evolution in contemporary economically developed society where infants may be more isolated than they were in our past.

Our mammalian relatives know how to care for their young—look at films of bears and wolves with their offspring. They may lack language, but they deeply understand. Sarah shares the same mammalian evolutionary heritage. Without the billions of emotionally and attuned Sarahs across the globe, our children would be psychologically crippled.

The famous psychoanalyst, Carl Jung, divided human sensibilities into thinking, feeling, sensation, and intuition. I know of no clear understanding of these terms. These sensibilities are parts of how we manage to live our lives forward, deciding, acting, doing, without knowing the future. They are, somehow, the armamentarium, the tools, evolution has equipped us with to carry out practical action in the persistent face of not knowing.

There are terms in our language such as metaphor. I have often found myself wondering why and how we use metaphor. Might part of the answer be that we must somehow span from what we know to what we cannot know, yet must somehow orient ourselves for practical action? Do we use metaphors in part to evoke emotions or intuitions? I have described the struggle to find a theory of quantum gravity. Is the probing process fully rational? Paul Dirac, a famous physicist and close friend of Carl Jung, once said that the deepest pleasure in science arose from finding an interpretation for a deeply held image. As an undergraduate in the 1950s, I looked in the Dartmouth bookstore window at the new publications and knew that someday I would write a book filled with connected arrows and dots, with some complex mysterious structure, something like a cubist painting, hinting order and understanding almost visible—but not quite. Dots with arrows connecting them have dominated an amazing part of my professional life. So has the search for the order and understanding that many of us undertake in our daily quests as scientists, scholars, in business, law, the arts. What is this swirl of imagery and invention? How many of us have experienced the creative capacities of our own unconscious minds?

I know scientists of enormous intuitive capacity, others, equally brilliant, without that capacity—at least as I assess it. Is intuition merely an almost precognitive rapid rationality? Or is it something else, a spanning of the unknowable by metaphoric images or something else? I do not know the real answers to these questions. But they evoke at least part of what we do not yet know about ourselves and how we live our integrated humanity every day.

How do we act when we cannot know? What possible means do we have to carry on with life, or "living it forward," as Kierkegaard said? Yet we do. Like the fiber-optic companies that went bankrupt when satellite communication was invented, we often fail, but we struggle to move on. The manufacturers of steam cars lost out to the gasoline engine. Pressure from the gasoline and automobile companies for road construction in the United States has limited the growth of the rail industry. Nietzsche said in this regard, "We live as if we knew." Yes. We live forward into mystery. But how?

First, we bring to bear all our animal past, all that evolution has given us, reason, emotion, intuition, understanding, experience, metaphor, the capacity for invention, the capacity for seeing and seizing an opportunity in an adjacent possible. Second, as we saw in chapters 11 and 12, much of what we do when we intuit, feel, sense, understand, or act is nonalgorithmic. We do not yet know, for a non-algorithmic mind, what it might mean to construct a theory for its operation.

Let's begin to explore this with a simple case: chess. Chess is partially unlike the real world. In a chess game, the players, moves, and what counts as winning are all prescribed in the sense that the "legal" moves are predefined. For any board position, there is a finite set of legal board positions either of the two players can move to on her turn.

Chess clearly can be algorithmic: IBM's Big Blue, crunching through vast numbers of future positions as a result of learning the positions of Gary Kasparov, was able to beat Kasparov. We even know, in principle, how Big Blue learned to play chess. We wrote the programs that drove Big Blue. It is a different issue whether when Kasparov played Big Blue, Kasparov's actions were algorithmic. But chess is not like the real world in the sense that the set of moves is not defined in the real world, and our actions may well not be algorithmic at all.

Natural Rationality Is Bounded

I have had the good fortune of working with Vince Darley, who gained a "First" in mathematics at Cambridge University, became my student at the Santa Fe Institute and Harvard, and is now a major member of a company called EuroBios. Vince and Alexander Outkin, both of whom were my colleagues at BiosGroup, a company I founded to apply complexity theory to practical business problems, have just published a scientific book that is one result of our six years of research at BiosGroup. It is called *The NASDAQ Stock Market* and contains perhaps the most sophisticated model of an actual stock market in action that has been created.

I return to consider simple models of economic life, where, unlike the full real world, the set of "moves" is prestated, because it is, in the current context, relevant to how and why we live our lives without knowing what will happen. In particular, when we act with one another we often create a nonstationary—that is, ever changing—series of even very simple behaviors. We cannot readily predict the actions of the other. We really do not know. Even in this setting, it is not clear that we can always make sensible probability statements. The question, relevant to this chapter, will then become: how do we do it?

The simple model grew out of the economic theory of rational expectations described in chapter 11 and redescribed briefly here. The theory of competitive general equilibrium demonstrates that stocks have fundamental values, and that there should be very little trading on the stock market if we both know the value of a stock. But this does not fit the facts, for large volumes of trades occur on real stock markets such as NASDAQ on a regular basis.

Rational expectations grew out of an attempt to account for occurrence of speculative bubbles such as the tulip bubble in Holland a few centuries ago when speculation in cultured exotic tulips flourished and crashed, and stock market bubbles and crashes. The basic idea is that members of the economy have a shared set of beliefs about the behavior of the economy, they act on those beliefs, and this sustains the very economy in which they believe: thus, rational expectations. Short-term speculative bubbles are the result of self-consistent beliefs that themselves drive investment behaviors that sustain the bubbles in the short term. Note, by the way, that the theory does not explain why there is a high volume of trade, for if we share exactly the same expectations, why do we buy or sell a stock? Given a price, it is either above or below the value we believe is proper for the stock, so who will take the "other side" of a trade? But the deeper question is whether such rational expectations "equilibriums" are "stable." Since the equilibriums exist in a space of theories about one another as economic actors and the economy we help to create, it is not obvious that equilibriums will be stable to perturbations, for example, in our expectations, or theories, about one another. Our simple model below suggests that the answer is that such equilibriums are, in fact, not stable to such perturbations in our expectations about one another, and worse, that we generically generate such perturbations as we strive to succeed. If we are correct, rational-expectation theory will need to be modified. In part, we will see that we often cannot even make sensible probability statements.

Herb Simon, Nobel laureate in economics, wrote about bounded rationality, which was held to be in contrast to the hyperrational "economic man" of competitive general equilibrium who could calculate the probabilistic values of all possible dated contingent goods such that contracts could be formed and equilibrium prevail however the future unfolded. Simon said we are not hyperrational, and that we satisfice: we do not optimize, we settle for "good enough." In other words, rather than identifying the optimal solution, which can be so complex that even high-speed computers struggle to find it, we use rough-and-ready success criteria, we satisfice. Simon was entirely correct, but his idea has been difficult to put into mathematical terms. There is one way to be perfectly rational in a prestatable known world of all dated contingent goods. There are infinite ways to be less than perfectly rational in a knowable world.

Beyond even Simon's bounded rationality and satisficing in a *knowable world,* we have already seen our in principle inability to form a even a probability assessment for an event such as the invention of the tractor, or satellite communications. This, like Darwinian preadaptations, is the unknowable future, the *unknowable future world* in the adjacent possible that we do not know at all. Economists have a phrase for this, called Knightian uncertainty, but it has played little role in economics. The economists, however, have not yet recognized that this uncertainty is the in principle uncertainty and partial lawlessness of Darwinian preadaptations in the biosphere and the economy.

Vince Darley and I wanted to show, even before Knightian uncertainty, in a still *knowable world*, that rational expectations economic equilibriums were unstable to small alterations in the behavior of the economic agents, and that in a specific sense, rationality is bounded even if the set of possible actions was predefined—unlike the real world.

Here is the model: First, imagine you have the monthly price of wheat for twenty months. Suppose you want to predict the price of wheat for the twenty-first month. The mathematician Fourier showed that any wiggly curve on a blackboard could be approximated by a weighted sum of sine waves and cosine waves of different wavelengths, appropriately weighted. So we can approximate the price of wheat over the past twenty months by a Fourier series. Let's use Fourier's method to try to forecast the next month's price of wheat. Well, we could use the first Fourier mode, the average price of wheat for the first twenty months. In general, this will be a poor predictor of the twenty-first month's price of wheat. Or we could use a thousand Fourier modes or wavelengths. We would obtain approximately straight lines connecting adjacent month's wheat prices. This thousand-mode model does not predict the next month's price of wheat well, either. In the "average" mode case, we have underfit the data, hence cannot predict well. That is, we have not used all the information in the actual twenty monthly prices of wheat to help us predict the next month's price of wheat. In the thousand-mode case, we have overfit the data, fitting even the noise in the data, and cannot predict well, either.

This overfitting or underfitting issue is a very general one given a finite set of valid data. It is well known mathematically that to optimally fit the data, and neither overfit nor underfit, one needs to use a modest number of, here, Fourier modes, say four wavelengths. So to predict optimally, our theory must be of bounded complexity. It must use three to five Fourier modes, not one or one thousand. It is in this sense that Vince and I thought that natural rationality is bounded, for the rest of the model was invented to assure that one should always build a model of bounded complexity, a bounded number of Fourier modes which would, at best, approximate the past twenty months wheat prices, and only approximately predict the next month's price of wheat.

The next critical step in inventing the model was the further realization that predicting the next month's price of wheat assumed what the mathematicians call a "stationary time series." This means that over a very long time, the price of wheat will fluctuate around an unchanging average value. Further, the variability, or variance, of these fluctuations will remain the same.

Now expand our model to picture two people playing some game, say with payoffs. Let us bound their actions, just as actions are bounded in chess, but here to only two: raising or lowering their right arm. As the two players participate in the game, each creates a theory of the other's actions, and bases his own actions on that theory. This is the "rational expectations" step. In a steady self-consistent world, the actions and theories of both players would remain in place indefinitely. This would be a *rational expectations equilibrium,* in which the theories about the two person tiny economy model were self-consistently instantiated by the behaviors of the players holding those theories. But Vince and I knew that this would not occur.

As the game progresses, each of the player knows a longer time series concerning the actions of the other player—two hundred months of wheat prices, not twenty. In order to improve his own payoff, *each player is led to build a more complex model of the other player* than he had based on a shorter time series early in the game. Thus, each player uses more Fourier modes and, indeed, knowing more about the other player, can predict the other player more accurately.

But then what happens? Well, because each player has a more precise model of the other player's actions, that predicts more precisely the actions of the other player, *that very model, because of its increased precision, is more vulnerable to disconfirming evidence! That is, the more complex model is more fragile and easily broken by being disconfirmed.*

At some point, one of the players acts in a way not fitted by the very precise model of the other player. This second player can ignore the misfit, or else alter his model of the first player. Suppose the second player alters his model of the first player. But then the second player will behave differently, based on his new model of the first player. Because the second player behaves differently, the model of the first player no longer fits the behavior of the second player! The result is that the first player now must invent a new model of the second player. At this moment, it is critical to notice that even in this very simple game with only two actions, raising and lowering of arms, neither player can make a well formulated probability statement about what the other player will do. Even in this very simple case, neither player "knows," or can optimize, and lives in the face of mystery.

In short, we are led into a space of actions and models of one another that are coevolving and changing in time. In particular, when the models of one another change, each player's behaviors based on those models of the other player change, and players no longer have a long time series of stable behavior of each other. Each has two or twenty months' price of wheat with respect to the other, not two thousand months. So, as the game progresses, each player must invent a simpler model of the other compared to his earlier detailed model.

But this change in behavior means that the time series is not stationary at all. The mean and variance in the number of times each player raises her hand has probably changed from the previous "regime" of the game. This changing pattern of the time series is itself generated by the increasing fragility of ever more precise models. In turn, the models the players build of one another undergo an oscillation between simple, but robust models that yield self consistent behavior for some time—temporary rational expectations that then lead to increasingly complex models, until their increased fragility leads to disconfirmation and a new pattern of behavior of the players, creating the nonstationary behavior.

This is the essence of our theory. It implies that we will always have models of intermediate complexity of one another, say only three to five Fourier modes, growing to perhaps ten modes as models become more complex but fragile to disconfirmation, then collapsing back to three to five modes, and we create nonstationary time series whose nonstationarity means that we always have only a modestly long period of stable behavior of the other, hence can only build models of intermediate complexity to avoid over- or underfitting the period of stable data. The model works wonderfully.

This model also predicts periods of high variability in behavior and periods of low variability in stock exchanges. Just such a period of high and low variability in stock exchanges per day is seen in the stock market itself and is called heteroscadisity, meaning just that there are periods of high and periods of low volatility. There, is, at present, no underlying theory for this heteroscadisity. We have provided the basis for such a theory. Note, by the way, that this theory *does* predict substantial trading volume, while rational expectations do not clearly do so. Indeed, as hinted, our simple model seems to imply that the economic model of rational-expectations market equilibrium is unstable and needs to be expanded to include the fact that we generically create nonstationary time series when we interact. Interesting, our little model seems to fit something well known by stock traders: apparent "rules" emerge to guide stock investing, but to be "valid" for only short periods of time. It is as if investors have short term theories which break down and yield new short-term theories. As the theories change, investment behavior changes, and the time series is nonstationary. I emphasize that this parallel does not prove that the model Vince and I developed correctly describes the market, but it allows us to arrive at what might be the start of an adequate theory. To my knowledge, no further work has been done along this line.

This model is important for a number of reasons. Note that here we have kept the actions of the players constant. They can only raise or lower their arms. In the broader real world, say in history and policy choices, or the economy, or our daily lives, the set of next actions may be open in fundamental ways.

In addition, due to the coevolution of models and actions of the two players, and the resulting nonstationary time series, it does not seem that the players can form well founded probability statements about the actions of one another. The time series keeps changing and all statistical features about the patterns of raising and lowering arms may change over time. At best, probability statements may make sense within an "epoch" of stationary behavior, but even this is unclear, for the models of one another keep changing and thus the time series may be nonstationary on any time scale. If so, probability statements, on either the frequency approach or Bayesian approach growing out of Laplace's N door problem discussed earlier to the meaning of probability, seem precluded. The players really do not know, but must act, and do so, even here, in the face of mystery.

Some of the cases of our practical action appear to be prestatable. But what about practical business opportunities? Bob Macdonald, Harvard MBA, and president of BiosGroup, explores, invents, and discovers new opportunities in business on a regular basis. In a way remarkably like the happenstance way affordances of Lego machines can be used together in insightful ways, Macdonald does the same thing in crafting business strategies. I have asked Macdonald how he makes business decisions, and it's clear that a decision involves emotions such as fear of certain possible consequences, experience based on past events and outcomes, which shows analogies and disanalogies to the present circumstances, and invention. Somehow we see similarities and dissimilarities between past experiences and the current situation. What is it to see analogies? We have already seen how problematic the choice of the "relevant" features to compare is, let alone our notion of "similarity." Is there some finite list of features that we compare? How do we delimit the list? What is our "distance" measure between the cases? How do we create it? On what grounds do we "accept" those similarities and dissimilarities as "reasonable" or adequate? Sometimes, based on these past experiences, we weigh these analogies and make a decision. Sometimes, we act in a different way: we invent an entirely new solution with which we have no previous experience to build from. As Nietzsche said, "We live as if we know." Living as if we know, whether or not we muddle through or brilliantly move along a clear path, is part of the creativity of human culture.

Although it's not always clear what the adjacent possible consists in, recognizing and seizing an opportunity in the adjacent possible is a phenomenon that we do all the time in our business, legal, scientific, artistic, and our interpersonal interactions. We see dimly, weigh what we cannot fully know, and yet somehow act. The attempts to construct a theory of quantum gravity are related to this idea in that it is so abundantly clear here that these scientists *do not even have a clear formulation* of what an answer might look like. The search seems non-algorithmic, as I have argued. How do these scientists search?

Years ago, the economist John Miller and I were examining a unifacial stone scraper I had found outside my home in Santa Fe. John suddenly held it up in front of him and kept pressing it vigorously. Immediately it was apparent to me that he was using it as a channel changer. Of course, we both burst out laughing. In moments we had a new theory for the collapse of the Anasazi civilization in northern New Mexico. It appears that in 1250 they invented the television channel changer, seven hundred years before the invention of the television. Religious insistence on continued manufacture of these channel changers led to the economic collapse of the Anasazi. We loved our theory, but the archeologists did not. But then, what would the archeologists who actually study the Anasazi collapse know? This story illustrates the nature of opportunity and the timely seizure of that opportunity. John Miller saw and seized the opportunity to make a joke with the unifacial stone scraper, and counted on my understanding immediately his vigorous channel changing pumping of the simple piece of rock. So much for the psychological theory of behaviorism, which could never cope with my spontaneous laughter as a reliable learned response to a stimulus situation.

At the frontier of the current economy, there are new opportunities. These range from technical inventions such as my channel changer, which takes the economic web into its adjacent possible and might almost be foreseeable, to the rigidity of the engine block which is a Darwinian preadaptation taking the economic web into its adjacent possible in a way that is probably not foreseeable, to all of the uses of screwdrivers, some of which may also enter and expand the economic web by creating new complements to those novel uses of the screw driver. It also includes new organizational forms. For example, discount retailing, such as Wal-Mart, has widely displaced earlier forms of retail businesses that can no longer compete. The invention of discount retailing also involved seeing and seizing an opportunity. Many retailers did not expect discount marketing to have such an impact on their businesses. Can we say ahead of time what all the novel complements and substitutes for a given good or service are, including both routine new goods and fundamental innovations? We thus return to whether, when, and how well we can prestate the evolution of the economy. Many think peak oil, that day when we will recover in one year the largest amount of oil (after which recovery rates will decrease), has happened or will soon happen. We have only vague ideas about what the consequences will be, from mild to catastrophic, to new energy sources, to the entire modification of Industrial and post-industrial society.

We see opportunities and take them, acting as if we knew, but we do not, and we do not know the long term consequences of what we do. Yet we wish to act wisely. What does it mean to act wisely when we cannot calculate the consequences? This will be part of our issue in considering a global ethic.

Sometimes it seems as if the range of choices is narrow enough and the opportunity so obvious that we rather imagine it could be algorithmic—precisely the framed cases where current cognitive scientists can succeed. Sometimes—as in quantum gravity—we have no idea what we are doing, the space we are exploring, the components that might be of use, or even clearly what we are hoping to explain—the very opportunity we are trying to seize.

Note that everything I have said about our flow, only partially knowing and understanding, into the economic or policy adjacent possible, with the evolving economy and history as a whole at stake, in general occurs in all of cultural evolution. Perhaps to a certain extent we can hope to guide what comes next with a shadowed vision of what may be possible. If for no other reason, we must try to invent a shared global ethic that will help us shape what we will deem to be an appropriate global civilization. For the first time in human history we have both the necessity to do so as our cultures are crushed together by globalization of commerce and by global communications, and the elements of the means to jointly partially shape what will become via global communications and international discussions at many governmental and non-governmental levels.

This persistent becoming of culture, science, the economy, knowing, doing, and inventing is the ever building result of ourselves as full humans. It is our very making of meaning in our lives. It is emergent. And it is as amazing, awesome, and as worthy of respect as the creative biosphere. As we see ourselves in a creative universe, biosphere, and culture, I hope that we will see ourselves in the world in a single framework of our entire humanity that spans all of human life, knowing, doing, understanding, and inventing. The word we need for how we live our lives is *faith*, bigger by far than knowing or reckoning. A committed courage to get on with life anyway. How to live a good life with faith and courage is at the core of philosophic traditions dating back to Greece, with Plato stating that we seek the

Good, the True, and the Beautiful. Philosopher Owen Flanagan, in *The Really Hard Problem*, pursues this philosophic tradition, asking how we make meaning in our lives in a set of meaning domains, a "meaning space" including the sciences, arts, law, politics, ethics, and spirituality. Flanagan is right. We make our meaning touching all these spaces of meaning. The existentialist insistence that we make our meaning in a meaningless universe by our choices and actions was a response to reductionism, in which the universe is meaningless. But life is part of the universe. In a newly envisioned universe, biosphere, and human culture of unending creativity, where life, agency, meaning, value, doing, and consciousness have emerged, and that we cocreate, we can now see ourselves, meaning-laden, as integral parts of emergent nature. Whether we believe in a Creator God, an Eastern tradition, or are secular humanists, we make the meaning of our lives, to live a good life, in all these ways. And we act without knowing everything. What should and can we do in the face of the ignorance that we confront? Our choice is between life and death. If we choose life, we must live with faith and courage, forward, unknowing. To do so is the mandate of life itself in a partially lawless, co-constructing universe, biotic, and human world.

In face of this unknowing, many find security in faith in God. We can also choose to face this unknown using our own full human responsibility, without appealing to a Creator God, even though we cannot know everything we need to know. On contemplation, there is something sublime in this action in the face of uncertainty. Our faith and courage are, in fact, sacred—they are our persistent choice for life itself.

15

THE TWO CULTURES

If the metaphysical poets of the sixteenth century were truly the first in the Western mind to split reason from the rest of human sensibilities, as T. S. Eliot claims, this split resonates today at the core of our entire culture. It is a breach we must seek to heal. As Abraham Lincoln told us, "A house divided against itself cannot stand." Humanity's house is divided. At least it is in Western culture and now much of the modern secular world. But the split between reason and the rest of our sensibilities might be much more widespread, even in other civilizations and among those of faith in the West. What would it mean to reassemble the diverse parts of the human house on a single foundation? In truth, we do not know. History is irreversible. We stand, for the first time, at a juncture in human culture where we have the communication tools by which we can jointly examine and consciously try to understand and heal this split, answer the need for and invention of a global ethic for a global—hopefully diverse and ever creative—ongoing civilization, and learn to live with at least one view of our God as the creativity in the universe, God, our own invention, God to orient and guide us in our humanity as we live our lives forward, with faith and courage. Inventing a global ethics, and reinventing the sacred for our times, can be the fashioning of how we can use the symbol of God as the creativity in the universe to shape our single and collective lives.

Does logic force us to take the cultural steps I have just discussed? No, we are not logically forced to do so. These steps are not some kind of cultural deductions from the new scientific worldview I discuss here. These cultural steps are, I hope, creative ones that the new scientific worldview that we seem to be happening upon invites us to take. The theme of invitation is an important one for the rest of this book. It is part of our ethical reasoning, which I'll return to later.

As I mentioned in the first chapter, C. P. Snow wrote of the split between reason and the rest of human sensibilities. He wrote brilliantly of the two cultures in 1959, the humanities, including art, literature, drama, and music, and science split apart. For Snow, the arts were the domain of high culture, and science remained a cultural laggard. Now, perhaps, the roles are reversed. Einstein or Shakespeare, we seem to be told, but not both in one framework. How torn we are. How mistaken we are. It is time to look afresh. If we cannot know all that will happen, if Sarah understood her infant son, if we make our human way into the unknown in practical action, then the arts and humanities are crafts where we explore these aspects of our human lives. Einstein and Shakespeare must live in a common framework, that in which the universe is beyond reductionism and harbors emergence and vast and ceaseless creativity as it explores the adjacent possible. Shakespeare's world is, in fact, part of an expanded vision of Einstein's world, beyond Einstein's classical physics determinism in general relativity and its wondrously successful reductionism. Shakespeare's world is one in which agents, and even actors, act. Story and narrative are part of how we understand such action.

The humanities and arts in the past three-and-a-half centuries of Western culture, and now much of global culture, live, sometimes in their own view of themselves, in the shadow of science. Since Newton's *Principia* replaced legal reasoning as the highest form of reason, science has largely come to be viewed as the preeminent self-correcting path to knowledge. Here is not the place to analyze the path of science itself, whether it is the steady accumulation of ever more knowledge, or the periods of "normal science" and scientific "revolutions" about which Thomas Kuhn wrote some forty years ago in *The Structure of Scientific Revolutions*, with its attendant debate about whether one can even translate from the pre- to the post-revolution concepts in that science, or some more nuanced view. Science has achieved a kind of unspoken hegemony with regard to access to "truth."

Newton berates poetry as "ingenious nonsense." This is one step in the schism between science and the arts. I believe this schism, insofar as it really exists, needs to be healed. In response to science, the arts have reacted strongly. Christian Bok, a prize-winning Canadian poet, examines the relation between poetry and science over the centuries in his book entitled *Pataphysics*. *Pataphysics* means a playful slap at physics itself. Bok's play is one contemporary response to science. But it still sees poetry as having to respond to science.

A colleague of mine who is a professor of English at a major Ivy League university told me of her deep sense of confusion about the very legitimacy of the humanities. She felt that much of Western humanities was in a postmodern disarray. When I discussed the ideas in this book with the head of English at my own university, we were both astonished to find ourselves near tears as we felt together the importance of the humanities, and the legitimacy of the sacred as that which we ourselves choose to hold sacred. So choosing will be a joint cultural process. Perhaps there can be a new sense of freedom in taking responsibility for living our meaning-laden lives without a supernatural Creator God.

If Bok is right, the humanities in the West truly have bounced off science for 350 years adopting a variety of stances towards science. In the West, early-eighteenth-century art celebrated science. Alexander Pope's admiration for Newton was unsurpassable. Part of the view at this time, attributed by Bok to Aikin, is that poetry must base itself in science, since "nothing can be really beautiful which has not truth for its basis." Then poetry must learn its truth directly from science. Poetry, in this view and time, becomes subservient to science.

By the time of Coleridge, in the early nineteenth century, poetry is trying to find an independent stance: "A poem . . . is opposed to . . . science, by proposing for its immediate object pleasure, not truth." Wordsworth tries to unite something of science and poetry, "the knowledge of both the Poet and the Man of science is pleasure." The differences between these two great English poets of that time attest to the diverse ways poets, and with them, other artists, have responded to science over the years.

Slightly later in the nineteenth century, the Romantics oppose science. Keats condemns Newton for the "cold philosophy" that must "conquer all mysteries by rule and line." Blake condemns Newton for the "reasonings like vast Serpents" that must hang their "iron scourges over Albion." By the late nineteenth century, poetry is seen by Matthew Arnold as a source of values to augment science and its world of fact without value. Where science finds the facts, art including poetry will evoke the values that will replace those authored by the transcendent God that Newton's success replaced with a Deistic God who wound up the universe and let it go to follow Newton's laws.

From the twentieth century to the present, we have seen the emergence of the surrealists, in clear reaction to science, Derrida, and deconstruction-ism, with its faith in hermeneutics with the ambiguities of language and meaning, and stances like that of Christian Bok himself, in which science at the limits may be taken, in a Thomas Kuhn sense, as a succession of concep-tual systems that cannot be intertranslated—Einstein's sense of simultaneity and time versus Newton's sense of simultaneity and time. And so, Bok seems to say, science and poetry are more similar in their language games than might have been supposed. Yet Bok himself, heritor of C. P. Snow, sees the purpose of poetry, including his own, as engendering surprise and appre-ciation.

I say to my friend, Christian Bok, "You underestimate your own poetry, you are part of a narrative tradition that is a hundred thousand or perhaps 5 million years old. Story, including poetry, music, and other arts, is how we know ourselves." If you doubt me, look at the Cro Magnon wall paint-ings from thirty thousand to fourteen thousand years ago in the caves in Lascaux, and elsewhere in southern France and in Spain. Are you *logically* compelled to sense their humanity? Of course not. Are you humanly com-pelled? I hope so.

In all these responses by this tradition of poets, it has been accepted that science is the overriding holder of the truth, even if reductionistic and value free. And science is seen as the highest form of reason, while reason is celebrated as the unswerving, preeminent goal of humanity. We echo Plato's philosopher king. It is not an accident that, for Plato, it is a philosopher, king of reason, who is king. But Plato himself was wiser than that: our human goals, he taught, are the pursuit of the True, the Good, and the Beautiful. As Flanagan writes in *The Really Hard Problem*, we find our human meaning is a space of meaning including not just science, but art, politics, ethics, and the spiritual.

Is this struggle to connect the arts, let alone the humanities, to science actually a necessary struggle? What if, as I contend, science cannot begin to tell us about the self-consistent co-construction and partially lawless evolution of the biosphere by Darwinian preadaptations, nor predict the evolution of technology as novel functionalities are found, nor predict the very co-construction and co-evolution of human cultures and our historicity themselves as we create them. What if rationality is only part of Jung's tetrad of how humans find their way in the world and science itself begins to tell us that reason alone is an insufficient guide to living our lives forward? Then perhaps we must reexamine and reintegrate the arts and humanities along with science, practical action, politics, ethics, and spirituality, as Flanagan writes.

In a world we cannot completely know but in which we must act, it becomes important to realize that *rhetoric*, the capacity to bring our wisdom to bear in uncertain situations of life, is as important as logic. Rhetoric is not verbal trickery, it is persuasion in the face of uncertainty with respect to the needs to act. It is not an accident that Aristotle's *Rhetoric* was written when Athens was a democracy, where persuasion in politics and practical action was required. But rhetoric is a fully legitimate part of the humanities and the rest of our human sensibilities.

Perhaps then poetry and poetic wisdom are right and real. When Lady Macbeth says, "Out, out damned spot," Shakespeare is revealing humanity to us. Lady Macbeth desperately wishes to erase evidence that her husband has killed. We know and understand her in her human situation. What Shakespeare is revealing is the truth about human actions, human emotions, human motive, human reasons. Indeed, Harold Bloom, a Shakespeare scholar, goes so far as to say that Shakespeare "invented" human nature. Perhaps it is fairer to say that Shakespeare showed us, in one short lifetime, more of the depths of humanity than any other playwright in the history of the world. Shakespeare walks among us down these centuries because he not only possessed dramatic talent, but because that talent teaches and reteaches us about ourselves. So is poetry only pleasure? Is it only astonishment and surprise? Absolutely not. Sublime poetry, sublime literature, is a lens through which to view ourselves, our lives, and our world. It shows us the truth.

And yet today it really is often science that has become the gold standard of truth. Because I am not a humanist I may overestimate, but friends in the humanities and arts too often tell me that their fields are kinds of second-class citizens. Humanists regularly tell me that they are pursuing the "soft sciences"—a disparaging term implying a kind of disdain. Yet I do not think science should be put on a pedestal above the humanities, and I say this as a practicing scientist who loves my craft with its own imaginative roots and practical action. We need to know, understand, appreciate, and live in our full world. When a humanist reads and discusses, say, the poetry of a Mexican poet of enormous vision and compressed expression, like that of John Donne at the beginning of this book, need we remotely say that this understanding must be science? Science cannot alone speak for the richness of life, meaning, and understanding. Of course we know this, but we do not integrate it with science as another inventive human cultural activity into a coherent vision of ourselves. Nor, at present, do we know what such a coherent vision might be. Must we live with this rift? At least a start would be to see science as a human activity, with its similarities to the arts, law, business, medicine, crafts, pastoring, all domains of diverse human creativity and actions in the face of uncertainty. There is our joint humanity, right in front of us. Flannagan is right, we make our meaning in the spaces of science, arts, politics, ethics, and the spiritual. Plato was right long ago: we seek the Good, the True, and the Beautiful.

Indeed, music, as the ancient Greeks knew, affects us in ways that no other stimuli can. We're beginning to understand that music is processed in different areas of the brain compared to other forms of auditory stimuli. If we say that we appreciate the music of Mozart, do we mean that aesthetics is less-than-factual knowledge, scientific or otherwise? Should we mean that aesthetics is less about being in the world and understanding that world than factual knowledge? But factual knowledge is just that, merely knowledge of facts, so far devoid of value, meaning, and life.

Music is part of our full humanity, part of Plato's the True, the Good, and the Beautiful. Music, like other arts, enlarges our world. If our tools, stick, stone, wheel, computer, enlarge our world or knowledge and practical action, art enlarges what we know and what we understand as well. To think of Shakespeare again, we know humanity in a different way thanks to his art. Perhaps "understanding" is, in fact, the right word. Thanks to Shakespeare, we understand humanity in an enlarged way. Only part of how we make our ways in our lives is by knowing the truth of statements of fact about the world. So is the old view that poetry is a path to truth correct? Yes: art is not just appreciation and surprise; it can be truth.

I find Mozart's *Requiem* sublime. It is death, God, loss, glory, agony. But these are as much a part of our lives, including our understanding human lives, as are statements about galaxies. Are you logically forced to find the *Requiem* sublime? Of course not. This is not a matter of logic at all. The experience of the *Requiem* is part of our full humanity, our being in the world. Being in the world is not merely cognitive, it is the full integration of all of our humanity, imagination, invention, thinking, feeling, intuition, sensation, our full emotional selves, and whatever else we bring to bear. We could not make our ways without all of these. And Mozart, like Bach and Vivaldi before him, and Beethoven and Brahms after him, was part of the creation of our culture, an evolution of music as an art form, as were the impressionists, and as is technological, legal, and moral evolution. If T. S. Eliot is right that the metaphysical poets split reason apart from our other human sensibilities, if the Greeks did not do so two thousand years earlier, then after three-and-a-half centuries of spectacular science, I say it is time that we heal the rift. Our lives depend upon it.

Frank Geary, Frank Lloyd Wright, and other architects conceived of and oversaw the construction of buildings that are magnificent. The Parthenon remains magnificent. The cave paintings in Lascaux fourteen centuries ago, Cezanne, Van Gogh, and Picasso show us our world in ways that expand our world, expand our understanding of our world and our being in the world. If you do not understand war, look at Picasso's *Guernica*. In it we see, fully see, how we love to kill one another.

"Truth is beauty and beauty is truth," wrote Keats. If we allow ourselves to embrace humanity, Keats shows us how foolish we have become in the long shadow of the science I love.

Living involves knowing, judging, understanding, doing, caring, attending, empathy, and compassion, whether science, business, the law, the humanities, the arts, sports, or other ways of going about our lives. If we cannot marvel in our own created, lived, meaningful, unforeseeable human culture, we are missing part of the sacred that we have created and we can instead celebrate. Just because science replaced legal reasoning as the model of rationality hundreds of years ago, we need no longer live with that view. Why limit ourselves, as magnificent as science is?

A brilliant young friend, Mariner Padwa, knows a number of ancient central Asian languages. I learn from him that historiographers sometimes doubt the legitimacy of their endeavors: soft science again. Only a narrowed view that the search for something like natural laws and the capacity to deduce from them is the real legitimate intellectual pursuit of a scholar can underlie this doubt. But how unfortunate this is. The cultural and historical past of humanity is just exactly how we have, as a species, created our world. A celebration of knowledge and understanding is available to us. If much of our understanding is retrospective, and rarely predictive, that is not a failure, it is a mark of the very unpredictable creativity we live and can view with awe. We come to understand. Why would we deny historiography full legitimacy on par with any science? It has that legitimacy and happily research money on these topics can be plentiful, as it should be.

A young historian I recently met is studying the Portuguese colonization of Brazil. Brazil lacked an adequate population to establish, defend, and maintain itself. People indigenous to Brazil were treated by the Portuguese as noncitizens. The Portuguese leader at the time saw and seized an opportunity: change Portuguese policy, accept the natives of Brazil as full Portuguese citizens, achieve thereby a population to sustain a new colony. The policy "move" worked. He is revered in Brazil today. What modes of knowledge, understanding, judgment, and invention went into his seeing the problem in the first place, finding a solution to it, seizing the opportunity, and making it happen? His policy changed history. His policy also changed the planet and the universe, for, among other consequences, the Amazon forest is today sadly being destroyed. Cultural evolution is creative, but no one would feel that all of cultural and historical evolution is defensible as part of Plato's Good. I hope they find alternatives via the global economic web.

I hope these few examples will help us reenvision ourselves in our lives, in our cultures as we further create them, and move somehow toward a global civilization of some form, hence as we create our world. Full legitimacy for this spawning cultural creativity, our own invention, is part of reinventing the sacred. It is part of seeing ourselves whole in the world, not only beyond reductionism, but beyond science itself. The two elements of culture need not be divided.

16

BROKEN BONES

In order to live in the world in a new way, if we can actually do so and have the will to explore and achieve what will be required, we need to understand ourselves as Pleistocene animals of the hominid lineage, split off from other members of the mammalian and vertebrate lineage. Human nature, evolved in the Pleistocene, will change, but evolution is a slow process. We are capable of astonishing achievements, but we are also capable of atrocities beyond description.

Given global resource limitations of water, probable peak oil production per day, global warming, population pressure, with the requirements to have a functioning, sustainable, evolving economy, these challenges that face the world today, and our old propensities to profit, power, and empire, we are and will remain hard pressed to find our way with wisdom. Jared Diamond's book *Collapse* chronicles past failures. The forces toward future failure are very large. Now, more than ever before, now when we know more, know our past failures, now when we are more a still diverse global community of Western, Islamic, Turkic, Confucian, Japanese, and other civilizations, in the widest communication in our full history, now it truly is up to us together. We cannot survive failure. We cannot afford culture wars along the boundaries of our diverse civilizations. Not knowing the future, we must act anyway.

The bones of our hominid ancestors are often broken. One, but not the only, interpretation of the students of these remains is that the broken bones sometimes reflect injuries and presumably warfare. Similar inter-clan and intertribal warfare is common today. Chimpanzees sometimes kill one another. And they do so brutally. Humans have brought to bear trillions of dollars on perfecting weapon systems to kill in the names of our God, our nation, or self-defense, often while we ignore the ravaging of others that we do behind our own backs. We hear that soldiers from the U.S. are killed every day in current Iraq. The number of Iraqis killed is poorly reported. The "other" has always been dehumanized throughout human history. We must face this in ourselves, with open eyes.

The past one hundred years have brought humanity two world wars in which tens of millions were killed, the genocide of Armenians at the hands of the Turks early in the twentieth century, the Holocaust, Stalin killing millions of his countrymen, Bosnia, Rwanda, Saddam Hussein, Pol Pot, and more.

Through one of my companies, BiosGroup, I have consulted for the U.S. Joint Chiefs of Staff. I have also gotten to know General Paul van Riper, who was third in command of the U.S. Marine Corps, and others on the Joint Chiefs of Staff. The officers I have worked with were all concerned with the asymmetric warfare that modern terrorism brings us. All sought to prevent war, but history shows us that war is often excused by a trumped-up atrocity or threatened atrocity said to have been committed by the "other."

We are capable of atrocity. We are capable of slashing attacks for national power—the Third Reich, Russia attacking Finland, probably the United States in Iraq. We are capable of slashing attacks for tribal conquest: Rwanda in the twentieth century. We are willing to kill in the name of God. The impulse to war and conquest will not change; it is part of our human heritage. Nor will our general impulse to power. Civilization and culture must change and triumph over our impulse to kill. Of course, this is easy to say, unknowably hard to do. The profound question, as a new type of global civilization emerges, is whether we will find the will and the way. I expect that none of us knows. Surely we should be as conscious of this as we can. Many with more skill, training, and experience than I can conceivably muster struggle with these issues. International scholars, lawyers, the military, national and international and nongovernmental organizations face these questions every day.

The reinvention of the sacred is our own choice of what we will hold sacred in an emergent universe exhibiting ceaseless creativity. Those who believe in a theistic, all-powerful, all-good Creator God confront the theological problem of evil. How can an omnipotent, omniscient loving God allow evil to occur in the world? Within the Abrahamic traditions, there are a variety of answers to the problem of evil. Recent Jesuit thought includes the concept that God outside of space and time, is a generator of the universe, but neither omniscient nor omnipotent. This is one answer to the problem of evil. If we are to reinvent the sacred with ourselves as part of the real world, with all its wondrous creativity around us, then we have to come to terms with the fact that evil happens at our own hands, let alone for causes beyond our control. There is nothing sacred about killing one or millions of people, in the names of God or otherwise. Thus, there can be no place for implacable religious fundamentalisms—including Christian, Islamic, and an emerging Hindu fundamentalism—in the civilization we must create. And precisely since fundamentalists have such difficulty allowing for the possibility of divergent views, or envisioning the modification of their own views, coevolution of our diverse cultures with fundamentalists across the planet is a challenge the rest of us do not yet know how to meet. If we fail to create enough of a sharable world view, beyond reductionism, filled with value and meaning, forever open and evolving and the best of our shared wisdom, the drive toward fundamentalisms as our diverse civilizations collide may outstrip our better selves. We do face a new, starker clash of civilizations, let alone power drives, political and economic. There is a still poorly stated global race underway between a retrenchment into our diverse civilizations that may lead to wars on the frontiers between these civilizations, and the creation of a sharable, but still varied world view in which we can live in common in Flannagan's "meaning space." I hope we are hovering on the verge of taking responsibility for our own sharable meaning, our own collective actions, our own evolving cultures, and our own use of the God we invent for ourselves. Ultimately, it is our collective choice whether we continue to break bones.

Since killing is part of our heritage, no emerging global civilization can afford to be even remotely utopian. We will need legitimized careful power, political, military, economic, to prevent, or limit, the bloodshed we are so willing to unleash. We will probably need such power for the indefinite future.

I offer one cautionary thought. World government seems at first a good idea for global legitimate power. But the best of constitutions can be subverted. I personally fear an overpowerful world government. The founding fathers of the United States rightly feared an overpowerful central government, and invented a federal system to balance that central power.

In addition, the large number of nongovernmental organizations, welling up from the grassroots, may also emerge as an additional balancing force. Paul Hawken discusses this grassroots emergence of NGOs in the face of perceived or real high-level international corruption of power in *Blessed Unrest*. He should be taken seriously.

Since we do not know and must act anyway, we must do so with humility. We do not, in fact, know. Humility necessarily invites tolerance. In the global ethic that we so badly need to construct, there is no room for intolerance. If we can find a new pathway to a reinvented sacred, and a wise coevolution of our traditions, perhaps intolerance can fade away.

17

ETHICS EVOLVING

In this chapter, I wish to discuss briefly the vast topic of ethics. For the reductionist, in a world of pure happenings, ethics is meaningless. But as we have seen, agency emerged in the universe, and with it came values, doing, and meaning. Fundamentally, ethics traces its roots to this emergence in the universe of values. But agency is part of life and its evolution, and that evolution, and as we have seen, cannot be derived from or reduced to physics.

I will argue that of the two major strands of modern Western ethics, *deontological,* or based on moral "law," and consequentialist, or based on the consequences of our choices, neither is adequate to our moral reasoning. Even biological evolution shows this. Certain brain lesions change our moral reasoning from relatively deontological to consequentialist. This fascinating fact implies that natural selection has operated to tune human morality. In place of these two major strands of modern ethics, I will suggest that Aristotle came closer to what we need. Moreover, our moral reasoning evolves with our civilizations; thus we need to try to understand what it might be to evolve our moral reasoning *wisely.* The evolution of the law, for example British common law, is a good model.

From the natural selective emergence of agency and values, hence "ought," and further natural evolution of higher social primates, we may understand how the rudiments of human morality evolved. We do not need a Creator God to author our morals. Our "moral sentiments" are partially evolved and partially derived from the histories of our civilizations. Yet many of us cling to the belief that without a Creator God, morality would crumble. Indeed, much of the resistance to evolution as fact and science is the fear that without God to create the universe and to author moral law, Western civilization itself would crumble. I empathize with this fear, but it is not well placed. Evolution is not the enemy of ethics but its first source.

If the reinvented sacred is our chosen aspects of the creativity in the universe itself that we deem worthy of being held sacred, then we must seek the roots and legitimacy of our morality here rather than with a Creator God. To a substantial extent, people across the world's civilizations share many of the same moral principles, such as a regard for the sanctity of life, despite a diversity of religious beliefs. If we can create a nonthreatening spiritual space that can be shared by those who believe in a Creator God and those who do not, then we can reason together about our morality and seek to convince one another without killing each other. We do not need moral fundamentalisms any more than we need religious fundamentalisms.

THE EVOLUTIONARY ORIGINS OF MORALITY

A wonderful experiment was carried out with Capuchin monkeys. The experiment consists of two monkeys in two cages facing one another but separated by a partition so neither can see the other. Adjacent to these two cages is a third cage in which a third monkey can observe both of the other two. The experimenter feeds one of these two apples, bananas, and so forth. The second monkey receives scraps. At some point, the observer monkey, well fed itself, is given extra food. What does this animal do? It gives the extra food to the monkey who received the scraps. These monkeys have evolved a sense of fairness.

Other animals exhibit a sense of fairness as well. Frans de Waal describes this in *Good Natured*. These experiments and others like them are part of a much wider study of animal emotions. Frans de Waal and other scholars have demonstrated that there are striking similarities between human emotions and the emotions of other animals such as the Capuchin monkeys. Not only do our emotions have evolutionary roots, our ethics do as well.

Frans de Waal lists a variety of tendencies and capacities, essential to our human morality, that are shared to some extent with other animals. These include sympathy related traits such as cognitive empathy, norm-related characteristics such as prescriptive social rules, reciprocity such as trading, and getting along such as community concern about good relationships.

It is a deep and legitimate question in biology how such a sense of fairness to others evolved. The evolutionary debate hinges on a central issue: if natural selection acts only at the level of the individual organism, favoring those with more offspring, why should natural selection evolve organisms that show altruism for nonrelatives, such as the Capuchin monkeys above? The old argument in evolutionary biology, based on pure individual selection, has been that it is most important to help your immediate children first, then your more distant relatives in proportion to their genetic relatedness to you. This is called kin selection. This form of kin selection will not by itself yield altruism to nonrelatives. But it may be a stepping stone to altruism to nonrelatives.

For some forty years, a minority group of biologists has been slowly gathering evidence for what is called group selection. Here the idea is that if there are different groups of organisms of the same species, it might arise that genes favoring altruism to nonrelatives in one group would give that group as a whole a selective advantage in propagating compared to groups not having the "altruism" genes. It has long been recognized by virtually all evolutionary biologists that there is nothing logically wrong with group selection as a theory. The issue has been whether it might be common enough to play a role in real evolution.

The recent discussions are slowly shifting in favor of a substantial, if still hard to quantify, role for group selection. Let us accept such a role.

Once group selection is effective, then natural selection for altruism, even for the evolution of the higher primate sense of justice or fairness, becomes perfectly reasonable. Experiments such as that with the Capuchin monkeys and the cases reported by de Waal increasingly suggest that group selection is real, and therefore, that the very first source of hominid morality is evolution itself. In short, far from being the enemy of morality, evolution has yielded at least part of human morality because morality has offered a selective advantage to groups. A side note of considerable importance concerns the evolution of cheaters in such groups—those who would accept altruism but do not offer it. Group selection theorists have developed models in which, given the capacity to *recognize* the cheaters, norms such as ostracism can be expected to evolve within the group. How powerful must ostracism and extrusion from the human clan have been fifty thousand years ago? Death awaited. Thus the emergence of a behavior such as ostracism led to selective forces within the group to sustain altruism among nonrelatives. In ways that we are just beginning to understand, some of our ethical behaviors evolved for very good group-selective reasons. Still, ethical and moral reasoning goes far beyond what can be accounted for by evolutionary arguments.

ANCIENT AND MODERN
WESTERN ETHICAL THEORIES

In the *Nicomachean Ethics*, Aristotle assumed the legitimacy of ethical discourse, which some twentieth-century philosophers have strongly doubted, and sought to understand our ethical reasoning. Ethics concerns human action, which in turn rests on ascription of reasons, intentions, and conscious awareness and understanding of the act undertaken, and the distinction Aristotle stressed between voluntary actions as opposed to those which are forced upon us. Notice that for Aristotle and for current courts of law, the notion of a morally responsible free will, discussed in chapter 13, is central. His example, you may recall, was the sea captain confronted by a storm that might sink his ship who decided to throw cargo overboard. Aristotle asks whether his choice and action was voluntary or forced, and decides it was voluntary.

Aristotle's ethics is sometimes characterized as virtue ethics. He did not feel there was some single, coherent set of moral "laws" from which we could consistently deduce all of morality. He recognized that people have widely differing values, and that life and moral reasoning are complex. I will return to something like the Aristotelian view below. From the perspective of this book, conscious awareness, moral responsibility for act undertaken, and fully human agency are all ontologically real, for their evolutionary emergence cannot be reduced to or derived from physics alone.

In Western culture, since the time of the Greeks and Romans, a concept of "natural law" was, until the Enlightenment, the keystone for moral reasoning. Natural law was, roughly, the view that there is order in the universe, an idea derived from Greek thought, and that right living and natural law meant to live in harmony with the order in the universe, an order to be reflected in human action, right action.

In the West, the Enlightenment, in particular the Scottish Enlightenment, challenged natural law. As I mentioned, contemporary Western ethics dates from the work of David Hume, who famously argued that one cannot logically deduce "ought" from "is." That is, claimed, Hume, from what is the case, for example that women give birth to infants, one cannot deduce that mothers ought to love their children. To think otherwise is called the naturalistic fallacy. All of Western ethics since Hume resonates and ricochets off his naturalistic fallacy.

Hume is only partially correct. In chapter 6, I discussed the emergence in the universe of agency, and with it, values, meaning, doing, action, and purpose. We have seen that biology cannot be reduced to physics. Agency, an expression of life, cannot, therefore, be reduced to physics. But physics is the language of pure fact, with no notion of value, purpose, function. The inability to reduce biology and agency to physics is what underlies Hume's claim that we cannot deduce "ought" from "is." He is right in that we cannot reduce agency and biology to physics. "Ought" and "is" require different forms of reasoning.

The ricochets from Hume have led many moral philosophers to wonder about the "truth status," and even the legitimacy, of moral reasoning. Here the question about the truth status of moral claims has been an attempt to see whether moral claims are true in the same sense as statements of value-free facts about the world. Of course, they are not. Values cannot be derived from physics, from what "is," alone. The twentieth century philosopher G. E. Moore developed the "emotive theory" of moral language, in which the statement "It is wrong to kill a human" is equivalent to "Ugh," an emotional expression of disgust. Here moral reasoning itself has become illegitimate.

But Moore is not right. Life, agency, values, and therefore "oughts" are real in the universe. With values, human reasoning, as well as emotions, intuitions, experience, and our other sensibilities, about values becomes legitimate, not mere emotive utterances, and not mere statements of fact, but instead reasoning about values. Reasoning about what we ought to do in concrete situations is a large part of how we orient ourselves in our real lives. Our moral reasoning is a genuine aspect of living our lives forward, and surely not reducible to the physics of Weinberg.

Since the Enlightenment in the West, and Hume's naturalistic fallacy, there have been two major secular strands of ideas concerning a coherent ethics. Perhaps the boldest and most intellectually beautiful is that of philosopher Immanuel Kant. In his influential work *The Critique of Pure Reason*, Kant paid no attention to practical judgment. Pure reason, argued Kant, can never give rise to "priorities"—our values in this book, his "ends," or purposes, for which we act. Here Kant echoes Hume's naturalistic fallacy. One cannot deduce "ought" from "is." Both these great philosophers are right. But given priorities, Kant argued, practical reason consists of linked if-then statements to achieve a valued end. Thus, to go to the store, one should follow such and such a route. But what of the priorities themselves, the "ends" for which we act? Here Kant sought an internal deontological or lawlike logic to guide us—his famous categorical imperative: "Act such that the maxim of your action can be universalized." Here is the canonical example: It is wrong to lie. Why? Because if you always tell the truth, the maxim of your action can be self-consistently universalized. If we all tell the truth consistently, there is no self-contraction. However, lying only works in a context where most people tell the truth. Lying, therefore, cannot be universalized. If we all told lies all the time, lying would not work to trick others. It would be a self-defeating collective behavior. The maxim of the action, the lie, cannot be self-consistently universalized.

Kant's categorical imperative is the most brilliant effort to find a self-consistent deontological logic for ethics of which I am aware. But do we think it suffices? It is a routine question to imagine the following situation: I am in my home. A killer, intent on killing my wife, enters and asks me if she is at home. Should I tell the truth in this specific context? Kant would say yes. Most of us would say no. Most of us would conclude from this that Kant's brilliant attempt at a logically consistent ethics fails. These kinds of examples have plagued the deontological strand of post-Humean moral philosophy. We seem to have, as Aristotle might have said, *no single moral law*, or set of moral axioms, from which all moral action can be derived.

The second major secular ethical tradition since the Enlightenment in the West is the Utilitarian theory, sometimes broadened to the consequentialist strand of moral philosophy, for the aim is to achieve consequences that are, in some sense, optimal. In utilitarian moral philosophy the central concept is to act for the greatest good of the greatest number. Bentham and Mill are the initial proponents of this theory. But does it suffice? Ignore for the moment how we quantify the amount of "good," or "utility," something brings to each person. Suppose a killer holds one hundred people hostage, and says that if we pick one person at random to be killed, the rest will be released. Bentham and Mill are constrained by their logic to find this morally permissible. Most of us would be morally repelled. In short, the greatest good to the greatest number is also not a universal moral principle.

Interestingly, recent neurological experiments illustrate the utilitarian theory and its weakness. A group of patients with damage to a small region in the ventromedial frontal cortex has been examined with respect to moral reasoning. These patients are otherwise normal by available criteria. If confronted with the hostage example, most people will say that it is morally wrong to kill the specific person singled out randomly. We would enter into a complex set of moral issues to decide what to do. Instead, those with the ventromedial lesion, which appears to impair the connection between our emotional system and our rational system, have not the slightest hesitation in agreeing that the specific person should be killed to save the rest. They have become good utilitarians. This clinical finding is pertinent to us, for it shows that a specific neural structure, if damaged, can alter our moral reasoning. This powerfully suggests that our emotional systems and our rationality are united in our moral reasoning. It further suggests that *this aspect of the union of our emotional and rational system in moral reasoning has evolved by natural selection.* Again, evolution is not the enemy of ethics but one of its bases. We have not evolved to be pure utilitarians. Our reason, emotions, and intuitive capacities all come into play in our moral reasoning. Importantly, neuroscience is beginning to understand some of these aspects of our behaviors.

There is a further difficulty with utilitarianism as our sole moral guide. We have seen repeatedly in this book that we often, perhaps typically, do not know the full consequences of what we do. Often, I have argued, we cannot even form coherent probability statements about future events. Thus, a profound practical limitation of utilitarianism is that we cannot calculate "the greatest good for the greatest number." We just do not know what the long term consequences of our actions will be. Our moral reasoning must, therefore, wrestle with this truth. We must do our moral best even when we do not know the consequences. The boundedness of our moral reasoning in this day of rapid technological change is particularly poignant.

Thus, there appears to be no set of self-consistent moral "laws" or axioms that can self-consistently guide all moral choices by a kind of deductive moral logic. The needs, values, aims, and rights that we have can come into conflict in specific moral settings. Thus moral reasoning enters the picture. (Perhaps, given Gödel's theorem for mathematics showing that not all mathematical truths given some axioms can be derived from those axioms, we should not be so surprised that we also fail in moral reasoning to find a self consistent set of moral laws, or axioms that can guide us to all moral truths.)

THERE IS NO SINGLE SUMMUM BONUM

Moral reasoning, it appears, is more like the Aristotlean picture than post-Enlightenment theories. A mathematical idea, called "global Pareto optimality," may well be of use here. Grant that we have overlapping but diverse moral values. Now imagine a set of possible alternative moral courses of action, a kind of "space" of moral policies. In this space, think of the moral "fitness" of one policy with respect to one moral aim as a "height." It is critical that among our moral aims are Kantian means to an end, instrumental oughts, and ends in themselves. *Between the moral ends in themselves, we have no way of assessing their relative moral importance.* Then the mathematical idea of an optimal solution among a set of values that cannot be compared relative to one another is that of a "globally Pareto optimal" set. Here is the idea. *A moral policy is globally Pareto optimal if any small change in that policy decreases the moral "fitness" of one end in itself with respect to other ends in themselves.* In short, if we have moral ends, and no means to compare their relative values, then global Pareto optimal moral policies are ones that cannot be changed without sacrificing one moral value at the price of another. *In general, there may be many global Pareto optimal policies, and, granted that we can have no measure of the relative values of our multiple moral ends, we can have no means to choose among the global Pareto optimal moral policies.* It would be deeply interesting to understand the extent to which our moral reasoning struggles to choose among diverse globally Pareto optimal moral policies. This struggle may, in part, emerge in the evolution of our system of laws and the roles of different precedents in specific situated cases, as I discuss below.

I think the view of moral reasoning as seeking global Pareto optimal moral policies is a useful one. *It replaces the Greek search for a single supreme "good," a summum bonum, with a search for at least one globally Pareto optimal moral policy in a concrete situation.* I would emphasize that this view, intuitively aligned with Aristotle's view that there is no single moral law, encompasses both the Deontological and Consequentialist views of morality, for all are moral ends in themselves, and part of our moral reasoning where global Pareto optimal moral policies are the natural "solutions." However, if there are, in general, more than one morally global Pareto optimal solutions among ends in themselves in any given moral situation, or over a wide set of such situations, then the Greek hope to find a *single* summum bonum will fail.

Yet this account does not reckon with the fact that our moral values change as our civilizations evolve. Thus, our moral reasoning evolves. I turn next to explore this issue, and will suggest that our moral ends in themselves can evolve wisely.

The Evolution of Morality

In his trenchant book, *Letter to a Christian Nation*, Sam Harris notes that one of the Ten Commandments is "Thou shalt not steal." At the time Moses engraved this on stone tablets on Mount Sinai, this law meant that one Hebrew should not steal the goods of his neighbor. But as Harris points out, those goods included the neighbor's slaves. So God's law obviously accepted slavery 4,500 years ago. We do not accept slavery now. Our morality evolves. In the early old testament, it was ten eyes for an eye, then an eye for an eye, then, with Jesus, love thy enemy. Our morality evolves. We know our morality evolves with our civilizations and history. It is foolish to think otherwise, to become locked into a moral stance that made sense in distant times. Fundamentalisms based on the ethics of long ago can become tyrannical. But then perhaps it is wise to seek models of how morality might evolve, but do so *wisely*.

As we consider the evolution of moral reasoning, we may wish to explore the possibility that we ought to take the evolution of law as our model in an emerging global civilization for how we might coevolve our moral traditions. The virtues are those of accumulated wisdom, the capacity to change as society changes and issues arise, and a tendency to repulse totalitarian ethics. Confronting the variety of situated moral issues, the law has evolved cross-cutting legal claims, often in the form of conflicting precedents, to be applied in concrete cases. These cross-cutting legal claims capture to some extent the cross-cutting arguments of our moral reasoning in real situations. I suspect that these cross-cutting legal claims partially reflect struggle between alternative, incommensurate, global Pareto optimal moral choices. It could be informative to see if this is so, and, in turn, how the evolution of law tracks the evolution of our ends in themselves.

In short, law is one refined expression of moral reasoning. For example, English common law, a magnificent institution of ongoing legal/moral reasoning applied to specific practical, concretely situated cases, has grown since King John granted the Magna Carta in the thirteenth century. I remain fascinated by the beauty and intelligence of this English institution that has evolved in an emergent way that has built on itself in unpredictable ways over centuries. It is emergent because the very subjects appropriate to adjudication have evolved over time, and the law with respect to these diverse issues has evolved as conflicting precedents have been used by diverse judges down the centuries. Thus common law has co-evolved its collective wisdom. The interwoven logic, precedent, case specificity, and conflicting interests that the law must adjudge are wondrous. The common law is a witness to life lived in its full complexity ever since 1215 A.D. Perhaps legal reasoning is as powerful as scientific reasoning.

Moreover, the common law can change dramatically in fully unexpected, unpredictable ways. During the period when Cromwell had replaced English royalty, the deposed king Charles sought to regain the throne by instigating two civil wars. He lost both. Cromwell wanted parliament to try the king. But under the common law at the time, the king was above the law and therefore could not be tried by parliament. Cromwell sought out dozens of barristers, and at last found a Welsh barrister, who I will call West, willing to take Cromwell's case to try Charles before Parliament. On the first day of the trial under the charge of tyranny, West rose, Charles at his side, and began to read the charge. Charles used his cane to rap West on the shoulder. "Hold!" exclaimed the king. West continued to read the charge. "Hold!" repeated the king. West continued to read the charge. "Hold, I say!!" said the king, hitting West so sharply on the shoulder that the silver tip of the king's cane fell free onto the floor. "Pick up the tip!" Charles ordered West. West continued to read the charge. "I said, pick up the tip!!" repeated Charles. West continued to read the charge. Frustrated and ignored, Charles bent over and picked up the silver tip to his cane.

The members of parliament gasped, for in bending over in Parliament, the king had carried out the symbolic act of bowing before the authority of Parliament. Therefore, the king had symbolically accepted that *he, too, fell under the law*. Thus, the king could be and was tried by parliament. Since that time, in England and the English speaking world, the king, president, or others of highest authority have fallen under the law. So King Charles's accidental act of bowing changed legal history in the most profound way. By what twists of history, accident, and accumulation of tradition, and politically driven legislation with its pork barrel, honor-bound ways, has the law evolved as an emergent living body that regulates our civilized lives? Who could predict its happenstance, wise, archaic, up-to-date evolution?

Neither morality, nor the law, is fixed. Both evolve. How do morality and the law evolve? We have some clues. In the British common law, precedent rules. Some findings by judges are hardly cited in further cases. Other findings become landmark decisions that are thereafter widely cited and therefore change the law in large ways. By these precedent citations in specific cases, the law evolves to adapt to an ever-changing culture. So, too, does our morality: abortion rights are rightly contentious, for they are morally complex, with strong moral arguments to be made on both sides.

As a scientist I have long wondered how complex self-propagating, co-constructing systems coevolve with no one in charge. Common law is an example. So is the economy. So is the biosphere. No one is in charge in any of these cases. I have long wanted to look at the distribution of precedent citations, wondering if most precedents are rarely cited, while some are cited very often. If a plot were made of the number of citings on an X axis and the number of instances of precedents cited once, twice, and so forth on the Y axis, we could obtain a distribution. If we plotted the logarithm of the number of citings on the X axis and the logarithm of the number of cases of one, two, and an increasing number of citings on the Y axis, would we obtain a straight line down to the right, bespeaking a power law? My intuition is that we would. If we found this, it would suggest that the common law evolves via rare alterations in precedent that deeply alter the law, and more commonly by modest alterations. If so, why?

We may find laws about the coevolutionary assembly of complex adaptive systems, ranging from multispecies ecosystems where just such power-law distributions of extinction events have been found, to economic systems, where my colleagues and I are predicting and finding the same kind of power-law distribution of firm extinction events, thus a prediction that matches observations, as noted in the chapter on the economy. If the same holds true for British common law, where precedents coevolve with one another such that the law, somehow, remains a coherent "whole," in a way analogous to species coevolving and communities somehow typically remaining coherent wholes, and to technologies in an economy where at any time most technologies are jointly consistent such that integrated economic activity occurs, we may find laws of coevolutionary assembly of the deepest interest. Note that if we find such laws, as wonderful as that will be, it is extremely unlikely that any such law, power law or otherwise, will predict the specifics of the evolution of the law, economy, or biosphere.

If there are power-law distributions related to the biosphere, the economy, and common law, perhaps due to self-organized criticality, there are no grounds whatsoever to believe that these can be reduced to physics. And note as well that such laws may govern how self-constructing nonequilibrium systems such as the biosphere, our economy, and our culture expand into their adjacent possibles.

The evolution of common law may be our best proxy for the evolution of our morality as our society changes. While we may never be able to predict the specific precedents that sharply alter the law, King Charles and the cane, we may be able to predict the statistics of the evolution of the law. As in the case of the economy, certainly the evolution of the law is not predictable, and almost certainly it is nonalgorithmic. The same considerations can apply to the evolution of our moral reasoning, where our moral "rules" and considerations coevolve with one another in complex ways, just as do species, technologies, and legal holdings.

Thus we come to a point of far greater depth than the simplicity that the Abrahamic God handed down in the ten commandments. Our morality evolves, and does so for good reason. It must adapt to changing culture. The U.S. Supreme Court deals with these very issues as part of its mandate for the legal system. We have no analogous institutions for the evolution of our moral reasoning. But that reasoning remains deep, vital, and vibrant. We must not, therefore, seek self-consistent moral axioms that hold forever and settle all moral questions self consistently. Bless Kant for trying. Rather, we must continue the conversation forever as our culture and its circumstances change. Our moral reasoning, like our law, is ceaselessly changing, ever creative. But like law, some of our moral considerations are of deep antiquity and presumably have not changed for good reason—partially the effects of group selection in evolution, partially a long understanding of their wisdom. To say that our morality evolves is not to invoke a blind moral relativism. Rather, it is to invite respect for past moral wisdom, a hesitancy to alter old moral holdings, with enough flexibility to adapt to new facts.

Our laws have largely evolved in independent civilizations. As we approach the advent of a new type of global civilization, my own hope is that we will continue to invent institutions from international courts to nongovernmental agencies that act across national boundaries, to help shape our evolving laws and our morality, partially embedded in those laws, across our traditions.

Further, the fact that our moral reasoning evolves is likely to be of major importance to us. As we'll see in the next chapter on global ethics, ethics is partially evolved and partially ours to persistently reinvent wisely. Doing so is attempting to develop a framework that allows humanity to unite with a reinvented sacred, and undergird the global civilization that we will create over time.

18

A GLOBAL ETHIC

We lack a global ethic. We must construct one to prepare for the global civilization that is upon us, hopefully persistently diverse and forever evolving, forever inventing, and partially guide the form it will take. I say "partially" because we cannot know all that will happen, including the consequences of our own action. Here, too, we live our lives forward, with courage and faith.

Such an ethic must embrace diverse cultures, civilizations, and traditions that span the globe. It must set a framework within which we can hope to orient ourselves. It must be of our own construction and choosing. It must also be open to wise evolution. A rigid ethical totalitarianism can be as blinding as any other fundamentalism.

Recently I was honored to be invited to a "charette," or small discussion group, at the Interdenominational Theological Center in Atlanta, Georgia, home to six African-American denominations. The ITC, driven by one of its trustees, Paula Gordon, aimed to link theology and ecology in a broad movement the ITC refers to as TheoEcology(sm). The announced—and wonderful—focus of the two-day meeting was on finding the principles for a new Eden. I would hope that even for those civilizations in which Eden is not a familiar concept, the search for a new Eden would still be welcome. I find in what this group came up with, the start of a statement of a global ethic.

Principles for a New Eden
Emerging from ethical imagination, drawing
Humanity toward:
a World
WHOLE
OPEN
Where
LIFE FLOURISHES,
We
REQUIRE JUSTICE, RESPECT AND HUMILITY
EMBRACE CHANGE
EMBODY STEWARDSHIP
TRANSFORM WASTE AND EXCESS
AFFIRM DIFFERENCES
RECONCILE WITH NATURE
CHAMPION THE COMMON-WEALTH AND COMMUNITY
And our Actions are Energized by
A Progressing DIALOGUE, grounded in neutral language
FEEDBACK
DIGNITY, WISDOM AND MULTIPLE CENTERS

A new Eden. It is a powerful image. In the Abrahamic tradition Adam and Eve ate of the tree of knowledge and for this original sin were driven from the first Eden. But knowledge is no sin. What we need is a new Eden that we envision, at least in part, and can gradually move towards. A global ethic will form the ground rules for this search, for search we will as a global civilization emerges, or be doomed to its haphazard birth without the benefit of our choices.

Where are we now? The secular of the world place our faith in love of family and friends, ethics, democracy, and something like the wisdom of the market, where we have become consumers. Those of various religious traditions have the varieties of those traditions and their moral wisdom. Thoughtful Islamic scholars decry, as does the Catholic church, the mass materialism of the West, in the broad sense of the West that now includes much of secular society. While that materialism sustains the global economy upon which we depend, we face planetary limitations—population, water, oil, and mass extinction—that will demand of us a change in our value system, our global economy, and our lives. We have few affirmed overarching sets of shared values across all or much of humanity to guide us. How shall we find or choose those values? How will those values reshape the very "utility functions" our global economy serves, so as to sustain the planet and afford meaningful work? How can we possibly evolve towards such an ethic? Talking and thinking about it as a global community, to the extent we can, is a step. Perhaps, as a global civilization emerges, a new global ethic will emerge as well, only partially examined as it emerges organically. We cannot, after all, fully know what we do. Yet if we move toward a global ethic, it is hard to believe that that very cultural co-construction will not be part of the co-construction of a global civilization.

One clue to that ethic comes from the attitude of aboriginal people who kill animals to live. Almost uniformly, such people invent rituals that give thanks to the spirit of the animal killed and send the spirit back to enrich and replenish that which was taken. Why would people behave this way? The only reason that seems to make sense is that we sense the sacred in life itself. All of life. If such a sense of sacredness is within us, perhaps it can be called forth again. The vast majority of us buy our food in the sanitized atmosphere of a grocery store. Our dinner meat is often shrink-wrapped agony. We are distanced from often horrible lives and the frightening death that lies behind our meal. In *Fast Food Nation*, Eric Schlosser describes the horrors of mass methods to raise and "process" animals for our benefit, and it is simply staggering. What, then, are our obligations? Organic markets may be part of the answer. To some extent, and at some cost, we do care, as we're seeing from the rising popularity of the organic movement in general.

If we reinvent the sacred to mean the wonder of the creativity in the universe, biosphere, human history, and culture, are we not inevitably invited to honor all of life and the planet that sustains it? As we unleash the vast extinction the accompanies our global ecological footprint, we are destroying the creativity in the biosphere that we should rightly honor. This diversity, we should say, is God's work—not a supernatural Creator God, but the natural God that is the creativity in nature. We are, in fact, one with all of life. We have evolved from a joint last common ancestor, perhaps 2 billion or more years ago, itself presumably evolved from the origin of life on earth capable of heritable variation and natural selection.

Can I logically "force" you to see the sacred in the creativity in nature and join in basing a global ethic on that sacredness? No. Hume was right. "Is" does not imply "ought," even if "ought" is emergent and natural given nonreducible life and agency. Even if the "is" is the creativity in the universe of which I write. No, I cannot logically force you. But I can invite you. The very creativity in the universe, the wholly liberating creativity in the universe we share and partially cocreate, can invite you, for that creativity is a vast freedom we have not known, since Newton, that we shared with the cosmos, the biosphere, and human life. Accepting that invitation, while recognizing the evil we do and that happens, may be wise for us all.

I believe that an emerging global ethic will be one that respects all of life and the planet. But this ethic flies in the face of old aspects of many Western religious traditions: the rational economic man maximizing his "utility" is the extension of Adam, for whom God created all the creatures for human benefit. How very kind of God to give humans dominion over all of the world—and how arrogant we are to believe that that dominion is ours after all. For it is not. We are of the world, it is not of us.

As if we had the wisdom to manage the task of such dominion even as we have become the most abundant large animal on the planet. Think of the consequences of preventing forest fires only to inadvertently stop the small fires that are more or less natural, and thereby allow an understory of plants to grow unabated in the forests such that forest fires, when they occur, are now devastatingly large. Our ecosystems are more subtle than we are. We often do not know what we do. If the biosphere is incredibly creative, it is also incredibly complex. So part of the global ethic that must evolve should include a good bit of Quaker thought: we know so little of the universe that humility is wise. Yet we invent new products and spew them forth—the ozone hole comes to mind. We almost always manage to carry out the "technologically sweet," and cannot think through the consequences to the sixth generation, a Native American view. We simply must embrace the fact that we cannot know the full consequences of what we do, yet we are impelled to act. Any global ethics we create must embrace our inevitable ignorance as we live our lives forward.

Yet, in this emerging global ethic that acknowledges the enormous complexity of our planet, its life, and our cultures, we still must make choices. Humility is wise, but we still have to act and make decisions without always knowing the outcome. Thus, a part of an emerging global ethic must include an attempt to understand how to be wise when we are so bounded in our wisdom, hence how can we marry humility with practical action?

Above Santa Fe, New Mexico, is the largest single stand of aspen in the United States. Each fall this part of the country is a golden-yellow glory. Several years ago, tent moths invaded, and many of us feared the vast stand would die. It did not. Tent moths lay waste to their environment, and then become self-limiting. Humans have been no different. It is, after all, the "best and highest" economic use of our resources. We can continue to exploit our resources, as the Venetians laid waste to the Dalmatian-coast forests to build ships and the Mayans overran their land. As Jared Diamond remarks in *Collapse: How Societies Choose to Fail or Succeed,* most civilizations have overrun their resources. As he says, what did the person on Easter Island who cut down the last tree think of that act? We will cut down the last tree as well, unless, as Diamond says, armed by history and somewhat greater wisdom, we forestall the inevitable overrunning of our resources and destruction of our planet. With global self-interest and a broadened morality we are invited to embrace all of life and the planet.

For the first time in human history, we have the possibility to think and act together, bounded though our rationality may be. We largely know we must achieve a sustainable economy. We largely know we need that economy to afford meaningful work. If we are to respect all life and the planet, then the tools we have to gradually reconfigure our economy are laws, economic incentives like carbon credits, and altering our value system, which, in turn, will alter the utility functions of consumers such that the economy can "make a profit" fulfilling those altered values. This is already happening with respect to energy utilization and with respect to farm animals. We are making progress and becoming aware.

A global ethic requires far more than a sense of oneness with all of life and a responsibility for a sustainable planet. It involves the crash of cultures into closer contact in the emerging global civilization. When the civilization of the West touched the Inuit civilization, that civilization was substantially destroyed. The Hispanics of northern New Mexico struggle to hold on to a way of life that is over three hundred years old as the Anglo culture flourishes. As the world grows smaller, more and more cultures and civilizations are moving closer together. Not surprisingly, as Huntington said, in the post–Cold War world we face a clash of civilizations, Western, Islamic, Turkish, Confucian, Russian, Persian, Hindu, and Japanese civilizations—some modern, some not. Not surprisingly, our identities increasingly lie in these civilizations and often in their religious heritages. Not surprisingly, the frontiers bounding these civilizations are the loci of war. If we cannot find common ground faster than the resulting fundamentalisms are rising up, we must fear more loci of war. The task of finding a common spiritual, ethical, and moral space to span the globe could not be more urgent.

This very book is part of the clash between secular and religious society, seeking a different path. It is profoundly true that fundamentalisms, religious and otherwise, are on the rise. Yet some Christian Evangelicals are embracing the need to combat global warming and save the planetary ecosystem. If, as we evolve toward a form of global civilization, we can find ways to not feel threatened in our ways of life, if we can find safe sacred spaces, if we can together build toward a shared global ethic, we can also hope to reach out to those whose retreat into fundamentalisms reflects fear, not hope, and offer new hope in joint action. And for those whose fundamentalism is based in hope and specific faiths, my wish is that seeing one sense of God as the chosen sacred in the creativity in nature can be a shared sacred space for us all.

We do not know how to achieve such a coevolution, but it may be possible. We actually have hints of successes. In the connected first world, much of the emerging global civilization is, in fact, evolving ever new shared forms, new mores, new cuisine, new lingo, new music, new business practices, new law, new technologies, new ways of being in the world. When these transformations occur, those of us who participate may feel some mixture of modest threat and amusement. I personally love the very idea of Chinese-Cuban cuisine available in New York City. If we can share novel cuisine, where the threat to our ways of life is low, can we find ways to share on broader fronts?

Standing before us, as part of an emerging global civilization and a global ethic, is the question of how to allow to blend and coevolve further the civilizations of Tibet, Singapore, Chile, Europe, Uzbekistan, China, Japan, North America, and the Middle East. What will emerge? Will we all end up eating McDonald's and speaking English only? Or will ever new forms and marriages of cultures continue to arise in rich variety? Chinese-Cuban cuisine writ large? I dearly hope for the latter, with the requisite tolerance and joy at the diversity that can continue to unfold. Who knows what we can jointly create? We have never become a diverse yet global civilization; humanity has not faced this challenge before. We really do not yet know how to succeed. But the stakes are enormous: war or peace.

A global ethic includes the ways we will find to balance the power aspects of our humanity, sometimes working for broad good, sometimes serving smaller interests. We have never solved these power issues. They are, after all, the stuff of empires and world history. There may be deep reasons that such power structures—national, regional, whole civilizations—will continue to pursue the familiar quest for power. That does not mean we cannot find means to restrain their worst aspects and harness the best. Interestingly, power is often tied into our now familiar autocatalytic wholes. Ecologist Robert Ulanowitz has pointed out to me that autocatalytic systems, molecular, economic, and cultural, tend to draw resources into themselves and compete with one another for shared resources. Thus, it is a historical fact, noted earlier, that U.S. automobile manufacturers bought up trolley companies across the United States and put them out of business for fear of competition for customers in need of transportation. In the United States, the military-industrial complex that President Eisenhower warned about almost half a century ago flourishes, and sometimes ravages in the name of profit. On the other hand,

Ulanowitz himself has found the evidence that mature ecosystems of obviously autocatalytic organisms find a balance in which the total energy flow times the diversity of energy flow is roughly maximized. If mature ecosystems can balance the power needs of diverse species that also need one another in the niches they create, we may be able to find cultural-economic-military analogues to ameliorate the unfettered power misuses we see and become more interwoven and both interdependent yet robust, flexible and adaptable, in ways that fulfill our inherent quests for power within a stable framework of a global civilization, a global ethic, and its constituting laws and mores.

Some of these issues have been around for as long as we can remember, such as power interests. Some are new, such as the onrush of civilizations together toward some form of a global civilization. Some issues are culture bound and have a limited bearing on the possibility of the full unity of our humanity in all our diversity. Some of these issues may find a new place in a reinvented sense of the sacred in the world. There may be wise and foolish ways for our diversities of interests to commingle. Democracy, a political means to achieve compromise among conflicting minority interests, is likely to be a deeply useful political process as a global civilization emerges. A global ethic truly is ours to create. It can only help move us together in this direction if we reinvent the sacred, invent a global ethic, come together, and gradually find reverence—meaning awe, wonder, orientation, and responsibility in the world we share. Can I logically force you? No. Can the creativity in the universe invite you? Oh, yes. Listen.

19

GOD AND REINVENTING THE SACRED

Seeking a new vision of the real world and our place in it has been a central aim of this book—to find common ground between science and religion so that we might collectively reinvent the sacred.

We are beyond reductionism: life, agency, meaning, value, and even consciousness and morality almost certainly arose naturally, and the evolution of the biosphere, economy, and human culture are stunningly creative often in ways that cannot be foretold, indeed in ways that appear to be partially lawless. The latter challenge to current science is radical. It runs starkly counter to almost four hundred years of belief that natural laws will be sufficient to explain what is real anywhere in the universe, a view I have called the Galilean spell. The new view of emergence and ceaseless creativity partially beyond natural law is truly a new scientific worldview in which science itself has limits. And science itself has found those very limits. In this partial lawlessness is not an abyss, but unparalleled freedom, unparalleled creativity. We can only understand the biosphere, economic evolution, and culture retroactively, from a historical perspective. Yet we must live our lives forward, into that which is only partially knowable. Then since reason truly is an insufficient guide, we truly must reunite our humanity. And if so, we truly need to reinvent the sacred for ourselves to guide our lives, based on the ultimate values we come to choose. At last, we must be fully responsible for ourselves, our lives, our actions, our values, our civilizations, the global civilization.

If these lines of discussion have merit and stand the test of scrutiny over time, we will transition to a new view of ourselves, our humanity, and this, our world that we partially cocreate. In this view, much of what we have sought from a supernatural God is the natural behavior of the emergent creativity in the universe. If one image can suffice, think that all that has happened for 3.8 billion years on our planet, to the best of our knowledge, is that the sun has shed light upon the Earth, and some other sources of free energy have been available, and all that lives around you has come into existence, all on its own. I find it impossible to realize this and not be stunned with reverence.

Yet I hope for more. I hope that what we have discussed in this book can be one of the seeds of a global conversation that will orient this generation and the next generations towards the construction of a global ethic, and help us to create a vision and reality of an emerging global civilization forever diverse, creative, and tolerant—a new Eden, a new Enlightenment. The vision we have discussed rests morality in our own hands, and rests the restraints on the evil we do with we who cause it.

The very notion of "reinventing" the sacred is likely to be sacrilegious to those who have faith in a Creator God or gods. In the view of those of faith, God exists and the sacred is an expression of His being and law. There is, then, nothing "invented" about the sacred. Thus, "reinventing the sacred" stands in danger of inciting angry reactions. As our diverse civilizations rush together in a world of global communication, business, and travel, and our ways of life seem—and often are—threatened, fundamentalisms flourish. The very notion that we might choose to reinvent the sacred may be too threatening to embrace, or may seem pointless to billions of people of faith, or equally to secular humanists: indeed, it is important to realize that for millions if not roughly a billion of those of us who do not believe in a Creator God, we the secular children of the Enlightenment often feel that the very words *sacred* and *God* are utterly corrupted. Many who feel this way are revulsed by the death wrought in the name of God, and the aggrandizing certainty of some religious fundamentalists. The same secularists rightly fear any "sacred" for fear that it can become totalitarian.

If we are to consider consciously reinventing the sacred, and a sense of God as the creativity in the universe, many are those who will be profoundly opposed, but for deeply different reasons. Conversely, with caution, I believe we need to find a global spiritual space that we can share across our diverse civilizations, in which the sacred becomes legitimate for us all, and in which we can find a natural sense of God that we can share to a substantial extent whatever our religious convictions.

What we talk about here is already not too distant from some Abrahamic theological thought. Some Jesuits who are also cosmologists look out at a universe billions of years old, with about 400 billion galaxies each with about 100 billion stars, and reason that life is likely to arise at unknown places in the universe, and hold to a theology in which God is a generator of all this vastness. In the view of these Jesuits, this God cannot know when or where life will arise. Thus, this theology already sees God as less than omniscient and omnipotent. This is a Generator God, outside of space and time, who does not know beforehand what will arise in the universe God has created. This view is remarkably close to that which I am discussing in this book. Neither God nor human beings know how the biosphere, the economy, and culture will evolve. Even if this God exists, reason remains an insufficient guide to action. Even if this God exists but cannot know, this God cannot reliably answer prayer. If, as I advocate, we rename God, not as the Generator of the universe, but as the creativity in the natural universe itself, the two views share a common core: we are responsible, not God. But the two views do differ in their most fundamental aspect. One sees a supernatural Generator God as the source of the vastness around us. The view I discuss, beyond reductionism, partially beyond natural law, sees nature itself as the generator of the vast creativity around us. Is not this new view, a view based on an expanded science, God enough? Is not nature itself creativity enough? What more do we really need of a God, if we also accept that we, at last, are responsible to the best of our forever-limited wisdom?

What we are discussing here is also similar in many ways to Buddhism, a wisdom tradition without a God, based on thousands of years of investigation into consciousness. Because Buddhism does not hold to a Creator God and is a way of life, with deep understanding of our emotional-rational-intuitive selves, I would hope that the rest of us have much to learn from its years of study of human consciousness. If we must reunderstand our humanity, wisdom suggests that we use all the resources we can find.

No matter what religion we are discussing in relation to reinventing the sacred, God is the most powerful symbol we have. Then dare we use the word *God* as we reinvent the sacred? To do so risks profound anger from those who do believe in a supernatural creator God. Yet our sense of God has evolved from the God of the Old Testament who argued with Abraham and Moses, to the God of the New Testament, to the deist God of the Western Enlightenment, to our theist Creator and other senses of God today. Do we use the word *God* meaning that God is the natural creativity in the universe? We are not forced to do so. It may not be wise. This use of the word *God* is open to angry misinterpretation, for we have reserved this word in the Abrahamic tradition to refer to the Creator God. How dare we use the word *God* to stand for the natural creativity in the universe? Yet I say yes, we can and should choose to do so, knowing full well that we make this choice. No other human symbol carries the power of the symbol, God. No other symbol carries millennia of awe and reverence. Our task is not merely to be true to any fixed historical sense of God, in which case the supernatural connotations of our symbol God would mean that we should not use the *God* word, for to do so would be willfully misleading. I do not intend to be misleading. Our task is to engage the whole of our humanity and to be wise. It may be *wise* to use the word *God*, knowing the dangers, to choose this ancient symbol of reverence and anneal to it a new, natural meaning. God is our name for the creativity in nature. Indeed, this potent symbol can help orient us in our lives. Using the word *God* to mean the creativity in nature can help bring to us the awe and reverence that creativity deserves. Can *I logically force* you to this sense of the sacred? No. But the vastness of nature, the wealth of invention in the bio-sphere and human historicity can *invite* you. If all that is, all around you, came into existence naturally, then, as Whitman said, look at a single blade of grass in wonder, it contains all that is.

If the new scientific worldview I have discussed is right, a radical view requiring careful examination, we do live in an emergent universe of unending creativity, breaking the Galilean spell that all is covered by sufficient natural law. We can experience this God in many places, for this God is real. This God is how our universe unfolds. This God is our own humanity. No, we do not have to use the *God* word, but it may be wise to do so to help orient our lives. This sense of God enlarges Western humanism for those who do not believe in a Creator God. It invites those who hold to a supernatural Creator God to sustain that faith, but to allow the creativity in the universe to be a further source of meaning and membership. I hope this sense of God and the sacred can be a safe, spiritual space we can all share.

For me, Notre Dame illustrates the sacred. In fact, it is built upon a sacred Druid site. When the Spaniards conquered much of the Americas, including Mexico and what is now New Mexico, they built churches, including the beautiful Santuario de Chimayo, on the holy sites of the Native Americans. The ancient Pueblo peoples constructed *sipapus*, which are holes in the ground that symbolize from where the underworld people first emerged and populated the earth. The sipapu remains inside the Santuario de Chimayo, a small depression with holy dirt, taken by pilgrims, and replaced each evening by the priest, from which I, too, have lifted sand. We always build our churches, such as Notre Dame, on the holy places of those who have come before in order to capture and transfer their awe and reverence to the new gods or God. It may be wise, as we assume responsibility for ourselves, to use the word *God* to mean the natural creativity that created us and all around us.

If we call the creativity in the universe, biosphere, and humanity God, we are claiming some aspects of these for ourselves as *sacred*. And we are doing so knowing that the abiotic universe is uncaring, that terrible things happen to good people, and that we do evil. If it may be wise to call the creativity in the universe, biosphere, and humanity God, we cannot know how this symbol and we will evolve as we live with it. This God may become precious to us in ways we cannot foresee. If we must live our lives forward, only partially knowing, with faith and courage as an injunction, this God may call to us as we step into mystery. The long history of life has given us tools to live in the face of mystery, tools that we only partially know we have, gifts of the creativity that we can now call God.

We can find common ground across our diverse traditions, religious and cultural, and come together toward a global civilization, rejoice in the creativity in our universe, our shared biosphere, and the civilizations we have created and will continue to co-create. We can find common ground as we seek a new understanding of humanity. Such a quest can serve to bring meaning, community, solace, reverence, spirituality, tolerance, and generosity to all of us. This is the task of generations, for it can be the next stage in the cultural, moral, and spiritual evolution of humanity. For the first time, we have the means to communicate and choose. Can we know what we will create together if we embark on such a discussion and quest? Of course not. How wonderful though—we have to invent it together. More, if we do not seek to reinvent our sacred together, our global retreat into fundamentalisms threatens continued lethality unopposed by a cultural evolution that can invite their willing participation. There is only one humane way forward—we must together reinvent a shared sacred and make it a safe space for all of us.

Can we reinvent the sacred? Think of all the gods and the God that humanity has cleaved to. Each has told its believers what is sacred. Whatever your own beliefs, what are we to make of these other beliefs? Either we all worship the same real and supernatural God in different names, or these gods and God are our own invented symbols. Against all of those who do believe in a Creator God, I hold that we have always created and needed this symbol. It is we who have told our gods and God what is sacred, and our gods or God have then told us what is sacred. It has always been us, down the millennia, talking to ourselves. Then let us talk to ourselves consciously, let us choose our own sacred with the best of our wisdom, always knowing that we cannot know.

The word *sacred* is, for many, tied inextricably with the concept of the divine, but in many instances it is used to express an immense respect or reverence. For example, the Pulitzer Prize–winning Kiowa poet Scott Momaday, whose phrase "We must reinvent the sacred" I have incorporated into the title of this book, speaks of an ancestor's warrior shield, captured by a Civil War–era Southern general and held in his family until rediscovered and returned to its Kiowa tribe of origin. Scott has explained the reverence with which this shield was received by his tribe, and how the shield was made sacred by the suffering that accompanied it. We know in our bones that some things are sacred.

If what is moral is not authored by a Creator God, but our own, partially evolved, sense, then there are no self-consistent axioms from which we can derive all moral behavior. Rather, there are convergent and conflicting moral views, and as thoughtful, reflective, mature people, we engage in moral reasoning with our full humanity about situations, laws, practices and ways of life. So, too, with the sacred, if *we* are the authors of what is to be held sacred then we will engage in mature reasoning about what is sacred. Might it be the case, then, that at this stage in human cultural evolution we are, at last, ready to assume responsibility for our own choices of what is to be sacred?

What if God and gods are our invention, our homes for our deepest spiritual nature? Is the Old Testament any less sacred if it is our invention, our language, our discourse? Is the King James Bible any the less miraculous if it is the writing of humans, not the transcription of God's words? Are the Shinto shrines surrounding Kyoto any less sacred if they are a fully human invention? Are Notre Dame and the soaring Gothic cathedrals of Europe, the Dome of the Rock in Jerusalem, the Wailing Wall remains of the second Temple in Jerusalem less sacred if seen as human expressions of our own spirituality, ourselves speaking to ourselves, invoking that spirituality in architecture? These human cultural inventions are sacred.

It is up to us to choose what we will build together across our traditions, across science and the rest of our cultures and histories. In reinventing the sacred, with God our chosen symbol for the creativity in the universe including our own capacity for our inventions of religion, I believe we can, at last, take responsibility for what we call sacred, and thus treat as sacred.

The God we discuss, then, might be God as the unfolding of nature itself. Such a God is not far from the God of Spinoza. But unlike Spinoza, I wish to say that we are partially beyond the Galilean belief that all is law governed. Indeed, the creativity in nature is that which is partially not law governed. But that very creativity is enabled by that which is law governed. Thus, we may wish to broaden our sense of God from the creativity in nature to all of nature, law governed and partially beyond natural law. Then all the unfolding of nature is God, a fully natural God. And such a natural God is not far from an old idea of God *in* nature, an immanent God, found in the unfolding of nature. Whether God is immanent in nature's magnificent unfolding and becoming, or nature itself in its magnificent unfolding and persistent becoming is God, there is an essential difference. We do not need to believe in or have faith in God as the unfolding of nature. This God is real. The split between reason and faith is healed. The split between reason and the rest of our humanity is healed. This that we discuss is a science, a world view, and a God with which we can live our lives forward forever into mystery.

And there is a place for devotion in this view of God. The planet and all of its life are worthy of our devotion in this reinvented sacred and global ethic. There is a place for spirituality as well. It is ours. We need only claim it. The ancient religious traditions in the world have accumulated profound wisdom. The same traditions pastor to their members, create community, give lives meaning. If our goal is to coevolve our traditions, secular and religious, without losing their roots, or their wisdom, towards a shared spirituality, the same pastoring and community is needed. And we need ritual, it is part of our feeling and knowing our membership. We will have to discover how to pastor a global flock with still divergent beliefs as we come together, and do so safely.

If we do these things, they must be done gently and with empathy. Fundamentalists need to be invited to consider a shared vision as well. Perhaps those faiths will remain unchanged. Perhaps they will slowly evolve together with a natural God. It will be a long cultural journey, should we undertake it. But it holds the promise of enriching us all as we discover, and choose, what we wish to hold sacred. We may never fully agree, and we need never fully agree, but we can begin this cultural global journey by sharing the honest conversation and the journey with care for one another.

At last, there is the possibility of finding our way together, with a shared value system that we jointly create—our chosen global ethic, our chosen spirituality, and our chosen sacred. Let us go forth and find a global ethic and reinvent the sacred for our planet, for all life, and for ourselves.

ACKNOWLEDGMENTS

More than is usually the case, this book owes a very large debt to the many colleagues, students, and critics alike over the years since 1965. In a sense, this book is a gathering together of themes that have pervaded my intellectual and personal life. Those I will thank, more than is usually the case, are not responsible for my mistakes. I am confident that all my friends, critics, and colleagues would disagree with portions of what I say. In no specific order, I thank Philip Anderson, Brian Goodwin, Jack Cowan, Kenneth Arrow, Per Bak, Christian Bok, David Foster, Amy Brock, Marina Chicurel, Brian Arthur, Bill Macready, Jose Lobo, Karl Shell, Philip Auerswald, Kelly John Rose, Joshua Socolar, Ilya Shmulevich, Roberto Serra, Pauli Ramo, Juha Kessli, Olli Yli-Harja, Ingemar Ernberg, Max Aldana, Philip Clayton, Terrance Deacon, Michael Silberstein, Lewis Wolpert, Katherine Piel, Philip Stamp, Bill Unruh, Dennis Salahub, Nathan Babcock, Hilary Carteret, Mircea Andrecut, John Maynard Smith, Philip Kitcher, David Alpert, Marc Ereshevsky, Murray Gell-Mann, Joseph Traub, Pamela McCorduck, Cormac McCarthy, Owen Flanagan, Daniel Dennet, Lee Smolin, Robert McCue, Gordon Kaufman, Paula Gordon, Bill Russell, Russell Taylor, Randy Goebel, Barry Sanders, David Hobill, Sui Huang, Vince Darley, Robert Macdonald, Sarah Macdonald, Robert Este, Christopher Langton, Sergei Noskov, Gordon Chua, Hervé Zwirn, Norman Packard, Doyne Farmer, Leon Glass, Richard Bagley, Walter Fontana, Peter Wills, and Peter

Grassberger. This group is eclectic, with physicists, chemists, biologists, philosophers, economists, business colleagues, theologians, and people in the media concerned with serious discourse. I am lucky to know and have been influenced by them all.

I thank my literary agents, John Brockman and Max Brockman, for their care and concern. And I deeply thank William Frucht and Amanda Moon, my editors at Basic Books, who have helped, as do fine editors, in seeking the statue in the hacked marble an author produces.

Finally, I thank my wife, Elizabeth; our son, Ethan; and our daughter, lost now these twenty-one years, Merit, who with my surviving family remains ever loved.

Stuart Kauffman
Calgary, Alberta

NOTES

CHAPTER 1

1   *Beyond Reductionism: Reinventing the Sacred*: The genesis of this book, and its very title, derive from what was for me a life-transforming, small, and quite wonderful conference in 1992 just north of Santa Fe, New Mexico, in Nambe. The Gihon Foundation, directed by Michael Nesmith, had set itself the task of organizing small, biannual meetings of three to five "thinkers" to ask, what is the most important question confronting mankind? One can find a certain amusement in the presumption that any three to five people could possibly say anything useful. We met for two-and-a-half days at Nesmith's beautiful adobe ranch—four of us: myself; Lee Cullum and Walter Shapiro, fine journalists; and a magical mountain of a man, Scott Momaday, Pulitzer Prize–winning Kiowa poet. Scott, perhaps six feet seven inches, some 270 pounds, bass voice, fixed us in his gaze and said, "The most important task confronting mankind is to reinvent the sacred." I was stunned. Trained as a doctor and scientist, even with a background in philosophy, it was beyond my ken to use the word *sacred*. The topic was outside the pale of my view of informed conversation. And I was instantaneously convinced that Scott was right. In the odd ways in which our lives intermingle in untellable ways, this entire book derives from that meeting. The foundation required us as a foursome by to write a position statement as to our view of the most important problem confronting mankind. We wrote, roughly, that a global civilization was emerging, that we were entering its early "heroic" age, when a new transnational mythic structure could be created to sustain and guide that global civilization, that we could expect fear and fundamentalist retrenchment as older civilizations were challenged, and that reinventing the sacred was part of easing the emergence of a global civilization. Despite the improbability of that Gihon meeting, I think we were

right. This book is my own attempt to fulfill our position statement. Perhaps, here, in print, I can take a moment to thank Michael Nesmith, the Gihon Foundation, Lee, Walter, and Scott.

3   *It is emergent in two senses:* Some of the material in this book was published before in an article with philosopher Philip Clayton. P. Clayton, and S. A. Kauffman, "On Emergence, Agency, and Organization," *Biology and Philosophy* 21 (2006): 501–21. Yet more of this book echoes my third book: S. A. Kauffman, *Investigations* (New York: Oxford University Press, 2000).

5   *a "natural law" is a compact description beforehand of the regularities of a process:* I am relying on my memory of conversations with Gell-Mann a decade ago. Murray may not have said, "beforehand." But in that case, his statement is more general, "A natural law is a compact description of the regularities of a process." That more general statement concerning natural laws covers both the availability of the compact description *beforehand,* as I have attributed to Gell-Mann, and will do throughout this book, as well as during and after the process. Therefore, if I misremember Gell-Mann's precise wording, his more general case includes my attribution "beforehand." If it were not the case that the more general statement covered times before, during, and after the process described compactly by the natural law, we would face the oddity of a law *not available beforehand, that only became a natural law after the event or process.*

CHAPTER 2

12   *A central feature of Newton's laws is that they are deterministic:* It seems worth at least noticing a tension between philosopher David Hume's conception of causality and the determinism of Newton's laws, Einstein's general relativity laws, and Schrödinger's deterministic wave equation laws for quantum mechanics. Hume argued that we reason from the constant conjunction of events, for example, two billiard balls, one hitting the other with the alterations in their motions, to causality and with it to determinism. Hume argued that this step was not warranted. All we actually see, he said, is the *constant conjunction* of the motions of the two balls over, perhaps, many instances. What warranted the claim that in the future the balls would continue to behave the same way? Only, said, Hume, the "principle of induction." That is, he said, we infer by induction that the future will be like the past. But then, he asked his famous question: By what logic or reasoning do we validate the principle of induction itself? Only that the principle of induction has worked in the past. But this, noted Hume, is a circular argument. We cannot, he said, validate the principle of induction. Many have attempted to deal with Hume's "problem of induction," including philosopher Sir Karl Popper, who famously attempted to sidestep Hume by claiming that scientists never make inductions, but boldly come up with hypotheses that

they then seek to refute. Popper relied on refutation to keep science distinct from, say, astrology. In turn, philosopher W. V. O. Quine showed that Popper's criterion of vigorous attempts to disconfirm a hypothesis does not work. Despite Quine's brilliant counterargument, given in this book, innumerable scientists today, sitting on panels deciding whether to grant money for a specific research project, persist in asking if the investigator has proposed work to show that he or she is wrong. We scientists tend to live with philosophies of science that are decades out of date. At the same time, we tend to feel that philosophy of science is a useless enterprise. It is all wonderfully curious. Meanwhile, I would just note my own *very slight* Humean concern about the determinism in the fundamental equations of physics. Just because Newton taught us to write down deterministic equations and even confirm them wonderfully (quantum electrodynamics is confirmed to eleven decimal places), one can hear Hume muttering in the background. The smallest length and time scale is the Planck scale of 10 to the −33 meters, and 10 to the −43 seconds. Are we really sure any equations, Gell-Mann's compact descriptions of the regularities of the process, hold at the Planck scale? Of course, no one knows what laws, if any, apply at the Planck scale. This is the subject of quantum gravity, described later in the book. I will argue later in this book, in "Breaking the Galilean Spell," that I do not think such laws obtain for the evolution of the biosphere itself, human economics, or human history. If I am right, then not everything that unfolds in the universe is describable by natural law.

CHAPTER 5

49 *Leslie Orgel, an outstanding organic chemist:* A. R. Hill, L. E. Orgel, T. Wu, "The Limits of Template-Directed Synthesis with Nucleoside–5' Phosophor(2-Methyl)Imidasziolides," *Orig. Life Evol. Biosphere* 23 (1993): 285–90.

51 *David Bartel, a molecular biologist:* W. K. Johnston, P. J. Unrau, M. S. Lawrence, M. E. Glasner, D. P. Bartel, "RNA-catalyzed RNA Polymerization: Accurate and General RNA-Templated Primer Extension." *Science* 292 (2001): 1319–25.

54 *bilayered lipid membrane vesicles:* P. Walde, R. Wick, M. Fresta, Al Mangone, P. O. Luisi, "Autopoetic Self-Reproduction of Fatty-Acid Vesicles," *J. Am. Chem. Soc.* 116 (1994): 11649–54.

55 *Twenty years ago, Gunter von Kiedrowski:* S. Sievers and B. von Kiedrowski, "Self-Replication of Complementary Nucleotide-Based Oligomers." *Nature* 369 (1994): 221–4.

56 *The chemist Reza Ghadiri:* D. H. Lee, K. Severin, Y. Yokobayashi, and M. R. Ghadiri, "Emergence of Symbiosis in Peptide Self-Replication through a Hypercyclic Network." *Nature* 390 (1997): 591–94.

59   *My own theory of collectively autocatalytic sets:* S. A. Kauffman, "Autocatalytic Sets of Proteins," *J. Theor. Biol.* 119 (1986): 1–24; S. A. Kauffman, *Origins of Order: Self Organization and Selection in Evolution* (New York: Oxford University Press, 1993).

68   *Walter Fontana, Richard J. Bagley, and J. Doyne Farmer went further:* R. J. Bagley and J. D. Farmer, "Evolution of a Metabolism." In *Artificial Life II: A Proceedings Volume of the Santa Fe Institute Studies in the Sciences of Complexity,* C. G. Langton, J. D. Farmer, S. Ramussen, and C. Taylor, eds. (Reading, Mass.: Addison-Wesley, 1992). A similar idea was independently put forth by physicist Freeman Dyson. F. Dyson, *The Origins of Life* (New York: Cambridge University Press, 1999).

69   *chemical-reaction cycles in metabolism that are autocatalytic:* E. Smith and J. J. Morowitz, "Universality in Intermediary Metabolism," *Proc. Natl. Adad. Sci. USA* 101 (2004): 13168–73.

70   *Beyond this, Albert Eschenmoser:* Albert Eschenmosser, "The Search for Potentially Primordial Genetic Materials." Keynote lecture presented at the COST D27 meeting Chembiogenesis 2006, University of Barcelona, December 14–17, 2006.

70   *Robert Shapiro:* Shapiro, Robert, "A Simpler Origin for Life," *Sci. Am.* 296, no. 6 (2007): 46–53.

CHAPTER 6

72   *Agency, Value, and Meaning:* My book discusses these issues. S. A. Kauffman, *Investigations* (New York: Oxford University Press, 2000).

CHAPTER 7

88   *The Cycle of Work:* This topic is discussed in S. A. Kauffman, R. Logan, R. Este, R. Goebel, D. Hobill, and I. Shmulevich, "Propagating Organization of Process: An Enquiry." In *Biology and Philosophy*, in press.

CHAPTER 8

101   *Self-Organization:* This is a major topic of two of my books: *Origins of Order: Self Organization and Selection in Evolution* (New York: Oxford University Press, 1993); and *At Home in the Universe: The Search for Laws of Complexity* (New York: Oxford University Press, 1995).

114   *in a parameter space of* K *and* P: This issue is discussed in my book *Origins of Order: Self Organization and Selection in Evolution* (New York: Oxford University Press, 1993).

116 *we have recently shown:* A. Ribiero, J. Lloyd-Price, S. A. Kauffman, B. Samuelson, and J. E. S. Socolar, "Mutual Information in Random Boolean Models of Regulatory Networks." *Phys Rev. E.* In press.

117 *two groups deduced that the deletion:* R. Serra, M. Villani, A. Graudenzi, and S. A. Kauffman, "Why a Simple Model of Genetic Regulatory Networks Describes the Distribution of Avalanches in Gene Expression Data," *J. Theor. Biol.* 246, (2007): 449–60. P. Ramo, J. Kesseli, and O. Yli-Harja, "Perturbation Avalanches and Criticality in Gene Regulatory Networks," *J. Theor. Biol.* 242 (2006): 164–70.

118 *A second line of evidence:* I. Shmulevich, S. A. Kauffman, and M. Aldana, "Eukaryotic Cells Are Dynamically Ordered or Critical but not Chaotic." *Proc. Natl. Acad. Sci. USA* 102 (2005): 13439–44.

118 *A third line of evidence:* E. Balleza, M. Aldana, E. Alvarez-Buylla, A. Chaos, S. A. Kauffman, and I. Shmulevich, "Critical Dynamics in Genetic Regulatory Networks: Examples from Four Kingdoms." Submitted.

118 *The final line of evidence:* M. Nykter, N. D. Price, M. Aldana, S. Ramsey, S. A. Kauffman, L. Hood, O. Yli-Harja, and I. Shmulevich, "Gene Expression Dynamics in Macrophage Exhibit Criticality." *Proc. Natl. Acad. Sci. USA.* In press.

CHAPTER 9

120 *The Nonergodic Universe:* This topic is discussed in my book *Investigations.*

CHAPTER 10

129 *Breaking the Galilean Spell:* This topic is discussed in my book *Investigations.*

137 *infinite number of simulations, or even a second-order infinite number of simulations, to cover all possible quantum events in an infinite set of simulated biospheres:* It is a technical issue in quantum mechanics that if there were a quantum-wave function for the entire universe *and if nothing to measure it and induce classical behavior were present,* then one could, in principle, carry out a single vast calculation or simulation of the linear Schrödinger equation of the entire universe, and all possibilities, including all possible preadaptations and all possible human histories, would be in that vast computation in something called a Hilbert space. According to this view, one could calculate the probabilities of all possible organisms and preadaptations and histories of intelligent species. There are two deep problems with this extreme view, a modern form of reductionism. Most important, classical matter, such as the moon and tables and physicists' classical measuring apparatuses, does exist in the universe. When a quantum process is "measured" by interaction with such classical matter, such as leaving a track in a chip of mica, the single, vast computation of the wave function of the total universe is no longer valid.

This is because when the measurement event happens, in the Copenhagen interpretation of quantum mechanics, the wave function "collapses" to produce classical behavior, and what is called phase information is irreversibly lost. So the further time evolution of the Schrödinger equation must be reinitiated each time a measurement event occurs, and the *single*, vast quantum calculation in the vast Hilbert space cannot be carried out. In the real world, the physicist seems stuck with *an infinite number of simulations* that include all possible consequences of all possible quantum-measurement events. A second problem with this reductionist view is that "linear quantum mechanics" is needed to calculate all the probabilities in the Hilbert space. But phenomena such as Bose-Einstein condensation appear to be quantum-mechanical but nonlinear due to feedback of the quantum system on itself. If such nonlinearities exist in the quantum world, as many, but not all, physicists now think, then all bets are off with respect to calculating the probabilities of states in a Hilbert space. Then we cannot know the probabilities of all possible organisms and preadaptations or alternative human histories. We do not know the sample space ahead of time.

141 *we are a very long way from Weinbergian reduction:* It seems worthwhile to again point out the criticism of a classical physicist. Suppose we, like Laplace's demon, knew the positions and velocities (or momenta), which include the masses, of all the particles in the universe since the big bang. Ignore quantum mechanics and general relativity for the sake of argument. Then, said Laplace, the entire future and past of the universe could be calculated by this demon. This would, the reductionist classical physicist might say, include the emergence of the biosphere and of Darwinian preadaptations. My rejoinder is this: for sufficiently complex systems, such as all the particles in the universe in their $6N$-dimensional state space of positions and momenta, no computer within the universe could carry out the calculations. First, real computers have limited accuracy due to finite representation of numbers. This round-off error has cascading consequences, for example, in the chaotic dynamics of some classical systems of three or more variables. To the best of my knowledge, this chaotic dynamics is true of the solar system and is exemplified in the many body problems in physics. Examples of many body problems are many mutually gravitating masses under Newton's laws.

Second, even if this round-off problem could be overcome, no physical computer *within the universe*, at least that we can imagine, could carry out the calculation of Laplace's demon. Instead, the classical universe is, in itself, a perfect analog computer of itself, ignoring that this is an odd sense of the phrase *analog computer.* The positions and momenta of all particles are "represented" by this analog computer to perfect classical precision. Now, with digital computers, we know that there are many algorithms in which the unfolding of the algorithm cannot be prestated. All we can do is watch to see what happens. Analogously, *within this classical universe, my strong intuition is that we cannot predict the future detailed evolution of the entire universe, because this classical universe is a vast many body problems, hence almost certainly chaotic.* But then, while we can predict many things quite well, say, most of the planetary movements in the solar system for long periods, the specific evolution of this biosphere and the emergence of specific Darwinian preadaptations will almost certainly be knowable only as they happen, not beforehand. Like the algorithms noted above, we can only watch and see what happens. Then no natural law available in this universe is a compact description beforehand of the regularities of the processes of the evolution of the biosphere and its Darwinian preadaptations.

Thus, the biosphere is partially lawless, and self-consistently co-constructing. So, too, is the evolutionary emergence of an integrated physiology in multicelled organisms, where organs evolve new functionalities by preadaptation but must functionally fit coherently together with other organs whose functions (like the swim bladder) are also diversifying. Of course, it is natural selection that picks the successful lineages. These preadaptations are causally efficacious in the further evolution of the biosphere and of physiologies, and we have as yet no theory for this self-consistent, partially lawless, co-construction. We seem to be beyond Cartesian science.

143   *Intelligent Design:* This topic is thoughtfully discussed in *Intelligent Thought: Science Versus the Intelligent Design Movement*, edited by J. Brockman (New York: Vintage Press, 2006).

CHAPTER 11

150   *The Evolution of the Economy:* Further material relevant to this chapter can be found in E. D. Beinhocker, *The Origins of Wealth, Evolution, Complexity and the Radical Remaking of Economics* (Boston, Mass.: Harvard Business School Press, 2006). See also S. A. Kauffman, "The Evolution of Economic Webs." In *The Economy as a Complex Adaptive System* Vol. I, The Santa Fe Institute Series in the Sciences of Complexity, edited by P. W. Anderson and D. Pines (Reading, Mass.: Addison-Wesley, 1998). See also S. A. Kauffman, *Origins of Order: Self Organization and Selection in Evolution* (New York: Oxford University Press, 1993); *At Home in the Universe: The Search for Laws of Complexity* (New York: Oxford University Press, 1995); and *Investigations* (New York: Oxford University Press, 2000).

150   *the "econosphere" is a self-consistently co-constructing whole:* The economy, with human agents participating in it, truly is a co-constructing, partially unforeseeable, evolving system. It is just conceivable that the evolution of the economy is not only partially unforeseeable, but that human inventions are literally *partially lawless*. This possibility hangs on whether the operations of the human mind are "lawful." Philosopher of mind Donald Davidson thought that the

mind is what he called causally anomalous, and that no law connecting its behavior to neural behavior could be found. Further, in chapter 13, I will discuss the scientifically improbable but possible hypothesis that mind is partially quantum mechanical. As I discuss there, a quantum mind might depend upon the phenomenon of quantum decoherence, which may not, for sufficiently complex quantum systems and quantum environments in the universe, itself be describable by natural law. If that happens to be true, then, consistent with Davidson's view, human inventions may be partially lawless. In that remarkable case, the economy would be self-consistently co-constructing, even though central inventive and other steps in that co-construction were themselves partially beyond natural law.

160   *sustaining economic-technological ecosystem of complements:* Ecologist Robert Ulanowitz, who has also written about autocatalysis in ecosystems, has made the same point about autocatalytic systems, ecological or economic, commandeering resources. Personal communication, 2007.

167   *This phase transition has now been shown mathematically:* R. Hanel, S. A. Kauffman, and S. Thurner, "Phase Transition in Random Catalytic Networks." *Phys. Rev. E.* 72 (2005): 036117.

167   *Stefan Thurner and Rudi Hanel:* R. Hanel, S. A. Kauffman, and S. Thurner, "Towards a Physics of Evolution: Critical Diversity Dynamics at the Edges of Collapse and Bursts of Diversification," *Phys. Rev. E.* 76 (2007).

172   *In 1988 the physicists Per Bak, Chao Tang, and Kurt Wiesenfeld:* P. Bak, C. Tang, and K. Wiesenfeld, "Self Organized Criticality," *Phys. Rev. A.* 38 (1988): 364–372; S. A. Kauffman and S. Johnsen, "Coevolution to the Edge of Chaos: Coupled Fitness Landscapes, Poised States, and Coevolutionary Avalanches," *J. Theor. Biol.* 149 (1991): 467–505.

CHAPTER 12

179   *Simon's view fits with:* H. A. Simon, "What Is an 'Explanation' of Behavior?" In *Mind Readings, Introductory Selections on Cognitive Science,* edited by Paul Thagard, (Cambridge, Mass.: MIT Press, 2000).

179   *work by Warren McCulloch and Walter Pitts in 1943:* W. McCulloch and W. Pitts, "A Logical Calculus of Ideas Immanent in Nervous Activity," *Bulletin of Mathematical Biophysics* 5 (1943): 115–133.

179   *Two major strands have unfolded:* A useful summary of this material, whose logic I partially follow in some of this chapter, is in chapter 1 of E. F. Kelly and E. W. Kelly, *Irreducible Mind: Toward a Psychology for the Twenty-first Century* (Plymouth, U. K.: Rowman and Littlefield Publishers, 2007).

179   *such as that in human language and grammar usage:* The preeminent figure here is Chomsky. Chomsky, N. "Formal Properties of Grammars." In *Handbook of*

Mathematical Psychology vol. 2., edited by R. D. Luce, R. R. Bush, and E. Galanter (New York: Wiley, 1963), 232–418.

179   *introduction to these ideas, now called connectionism:* Connectionism is part of a vast topic that is sometimes called the computational theory of the mind, discussed in this and the next chapter. A major step in its direction was taken in *Parallel Distributed Processing: Explorations in the Microstructure of Cognitions*, 2 vols., edited by D. E. Rummelhart and J. McClelland (Cambridge, Mass.: MIT Press, 1986). One critique of the computational theory of mind is in (a review of *The Rediscovery of the Mind*, by J. Searle) T. Nagle, "The Mind Wins," *New York Review of Books*, March 4, 1993.

181   *computational theory models of the brain:* In the next chapter, where I again discuss the computational theory of the mind, I discuss the Chinese Room argument by philosopher John Searle, with references.

183   *theoretical neuroscientist Jack Cowan:* Cowan is a close colleague. Personal communication, 1999.

185   *Douglas Medin's research into human categorization:* D. L. Medin, "Concepts and Conceptual Structure." In *Mind Readings: Introductory Selections on Cognitive Science*, edited by Paul Thagard (Cambridge, Mass.: MIT Press, 2000).

188   *no finite list of affordances of the screwdriver:* Some computer scientists, however, are persuaded that machine learning will overcome this issue, for example, Alan Macworth, Department of Computer Science, University of British Columbia, personal communication, 2007. It is possible that Macworth is right, although I cannot see how. More generally, in the LegoWorld example notice that I used the Lego crane to pull the Lego platform loaded with Lego blocks across the Lego bridge over the toy river to a Lego house building site across the river. In doing so, I was using the crane in a nonstandard function, as a would-be tractor to haul the platform. But there seems no obvious way to prestate all possible uses of a set of objects in relation to one another and the rest of the universe for all possible purposes that might be of economic value. In real life, a real contractor might have made money using the crane to pull a real platform loaded with bricks across a bridge over a river to a building site, so this set of activities would have come into existence for a time in the real universe. I cannot see how to make this algorithmic at all. More, what occurred with the real if imagined contractor became part of the becoming of the universe. I would like to hope that something like Ulanowitz's Ascendancy, modified for cases of work, or activities, is maximized for partially lawless bio-spheres, supracritical economies, and civilizations as they evolve, diversify, suffer extinction events and gales of creative destruction, yet largely remain coherent as they build themselves. Perhaps something like this could become a "law" for partially lawless, self-consistently, co-constructing evolving far from equilibrium wholes.

192   *computations . . . are purely syntactic:* Philosopher John Searle makes similar points about the absence of semantics in a purely computational theory of the

mind. J. Searle, *The Rediscovery of the Mind* (MIT Press, Cambridge, Mass.: MIT Press, 1992); and J. Searle, *The Mystery of Consciousness* (New York: New York Review of Books, 1997).

192 *Claude Shannon's famous information theory:* C. E. Shannon and W. Weaver, *The Mathematical Theory of Communication* (Urbana, Ill.: University of Illinois Press, 1963). (Original work published in 1949).

194 *Lee Smolin's book* The Trouble with Physics: L. Smolin, *The Trouble with Physics: The Rise of String Theory, the Fall of a Science* (New York: Houghton Mifflin, 2006).

CHAPTER 13

197 *this hypothesis is highly controversial:* As I will make clear throughout this chapter, I believe the possibility that quantum events may play a role in consciousness is real and more defensible than most think. I will put forth the most cogent case I can for this possibility, which I believe warrants cautious investigation, perhaps in directions I shall indicate. Nevertheless, the "quantum mind hypothesis" remains scientifically unlikely at this point. The chapter is long, the rest of the book does not depend on it, and if the reader is not interested, he or she may choose to pass over it. An early discussion of these ideas can be found in the epilogue of my September 3, 1996, Santa Fe Institute preprint, "The Nature of Autonomous Agents and the Worlds They Mutually Create."

198 *First is the problem of free will:* Many neurobiologists, including Francis Crick in *The Astonishing Hypothesis: The Scientific Search for the Soul* (New York: Simon and Schuster, 1994), believe that free will is, in a sense, an illusion. This position notes that many neural events in our brains are not associated with conscious experience and may play a role in our decisions and actions as we experience them. Then the mind—on the mind-brain identity theory explained in more detail in this chapter—could be deterministic, yet we could have the conscious experience of free choice and action. At the end of this chapter, I will address free will "head on," so to speak, from a quantum-mind possibility. One of the issues to be addressed on either the familiar view above of a "classical" brain substrate for consciousness or a quantum mind-brain identity view, as I will discuss in this chapter, is mental causation—how the mind "acts on" matter.

200 *philosophers of mind, such as Patricia S. Churchland:* P. S. Churchland, "Consciousness: The Transmutation of a Concept," *Pacific Philosophical Quarterly* 64 (1983): 80–93. See also P. M. Churchland, *A Neurocomputational Perspective: The Nature of Mind and the Structure of Science* (Cambridge, Mass.: MIT Press, 1989).

200 *as the philosopher Owen Flanagan notes:* Owen Flanagan, "A Unified Theory of Consciousness," in *Mind Readings: Introductory Selections on Cognitive Science*, edited by Paul Thagard (Cambridge, Mass.: MIT Press, 2000).

201   *In the Chinese room argument, Searle asks:* J. Searle, "Minds, Brains, and Programs," *Behavioral and Brain Science* 3 (1980): 417–424.

202   *the neural correlates of consciousness:* Christof Koch, *The Quest for Consciousness: A Neurobiological Approach* (Englewood, Colo.: Roberts and Company, 2004). Koch's book is a fine summary of one very sensible line of attack in neurobiology to the problem of consciousness. Like all others, Koch offers no insight into what experience "is," the hard problem. For an important discussion of consciousness similar to that of Koch in spirit, see also G. Edleman and G. Tononi, *A Universe of Consciousness: How Matter Becomes Imagination* (New York: Basic Books, 2000).

203   *The binding problem:* See Christof Koch's *The Quest for Consciousness: A Neurobiological Approach.*

205   *quantum phase decoherence:* See a very accessible review of this substantial field by Philip Stamp. P. C. E. Stamp, "The Decoherence Puzzle." *Studies in the History and Philosophy of Modern Physics* 37 (2006): 467–97.

205   *Richard Feynman, to convey the central mystery of quantum mechanics:* See R. P. Feynman, *Lecture Notes in Physics* vol. 3 (New York: Addison-Wesley, 1965).

207   *decoherence is the favorite:* P. C. E. Stamp, "The Decoherence Puzzle." *Studies in the History and Philosophy of Modern Physics* 37 (2006): 467–97.

207   *a "bath" of quantum oscillators:* Ibid.

208   *decoherence approaches classicity for all practical purposes:* Ibid.

209   *but need not assure classical behavior with a quantum spin bath:* Ibid.

210   *the protein that holds it is the antenna protein:* H. Lee, Y. C. Cheng, G. R. Fleming, "Coherence Dynamics in Photosynthesis: Protein Protection of Excitonic Coherence," *Science* 316 (2007): 1462–5.

211   *Zurek points out:* Zurek is a close colleague. Personal communication, 1997. And see P. C. E. Stamp, "The Decoherence Puzzle" *Studies in the History and Philosophy of Modern Physics* 37 (2006): 467–97.

212   *reversion of partially decoherent "degrees of freedom" to fuller coherence:* See A. R. Calderbank, E. M. Rains, P. W. Shor, and N. J. A. Sloane, "Quantum Error Correction Via Codes over GF(4)," *IEEE Trans. Inform. Theory* 44 (1998): 13691387.

213   *quantum control:* There is a large body of literature on this topic. See A. Doherty, J. Doyle, H. Mabuchi, K., Jacobs, and S. Habib, "Robust Control in the Quantum Domain." Proceedings of the 39th IEEE Conference on Decision and Control (Sydney, December 2000), quant-phy/0105018. And S. Habib, K. Jacobs, and H. Mabuchi, "Quantum Feedback Control— How Can We Control Quantum Systems without Disturbing Them?" *Los Alamos* Science 27 (2002): 126–32.

215   *no account at all is taken of decoherence:* While current quantum chemistry largely ignores decoherence, and it seems worthwhile to integrate growing understanding of decoherence into quantum chemistry, it may be that such integration will yield only minor "corrections" to current quantum chemistry.

Alternatively, the alterations introduced by understanding decoherence in multiparticle systems such as proteins and the ordered water molecules next to and between them in cells may yield important new insights.

216   *a highly crowded, densely packed, and organized arrangement:* B. Alberts, et. al., *The Molecular Biology of the Cell,* 3rd edition, (New York: Garland Publishing, 1994).

216   *quantum coherent electron transfer within protein molecules:* A. de la Lande, S. Marti, O. Parisel, and V. Moliner, *J. Am. Chem. Soc.* 129 (2007): 11700–7.

216   *separated by beween 12 and 16 ångströms:* D. Beratan, J. Lin, I. A. Balabin, and D. N. Beratan, "The Nature of Aqueous Tunneling Pathways Between Electron-Transfer Proteins," *Science* 310 (2005): 1311–3. (Also, in note just above, see de la Lande et. al.)

217   *percolation theory, which is also well established:* D. Stauffer, *Introduction to Percolation Theory* (London: Tahlor and Francis, 1985).

220   *percolation theory approach to vast quantum–coherent webs:* This percolation theory approach is one possible realization to the discussion in the epilogue of my 1996 Santa Fe Institute preprint, "The Nature of Autonomous Agents and the Worlds They Mutually Create." There I explore the Bohm interpretation of quantum mechanics and its account of decoherence, and explore the idea of webbed loops of quantum coherence that sustain themselves through Bohm's "active quantum information channels," and persistently decohere through Bohm's "inactive quantum information channels." As far as I know from my physicist colleagues, Bohm's interpretation of quantum mechanics is consistent with the known phenomena of quantum mechanics.

222   *The Poised Quantum Mind Hypothesis Again and Philosophic Issues:* I thank philosopher Michael Silberstein, the University of Maryland, for long and helpful discussions. Silberstein has written extensively on emergence. See also P. Clayton, *Mind and Emergence: From Quantum to Consciousness* (Oxford, U.K.: Oxford University Press, 2004).

228   *can be identical to mind's intentions:* Consider the issue of free will in a conceivable quantum conscious mind that acts and can be morally responsible. We are said to be trapped. If the mind in question is deterministic, then it is not a free will, and not responsible. If it is probablistic chance, including quantum probabilistic randomness, then it may be acausal and "free," but not responsible. However, we have the hint of an opening wedge. The detailed way decoherence happens in a sufficiently complex and diverse quantum system decohering in a complex quantum environment and the universe, or a mixed classical and quantum system decohering in a mixed classical and quantum environment, may be different and unique in the history of the universe each time it occurs. Then no description of the regularities of the specific decoherence process would be available, hence the specific decoherence process would be partially lawless, or, otherwise stated, contingent. But no probability statement, on either a frequency interpretation of probability or the Laplacian N door "ignorance" basis, can be made. The sample space is not known. The same claim would seem to

apply to a sustained partially coherent quantum system as mind. Then the mind's behavior is neither deterministic nor probabilistic. This "wedge" may afford a responsible free will.

In a sense, this discussion bears a resemblance to chapter 7's discussion of propagating organization of processes in cells building classical boundary conditions that are constraints on the release of energy that does work to build further constraints on the release of energy that does more work to build more constraints on the release of energy. Quantum motors are known, in theory, to be able to extract work from coherence. If decoherence of the quantum world alters the classical world by becoming part of the classical world, but the classical world alters the boundary conditions of the quantum processes, as may just happen in the brain, it seems at least fair to say that a lot remains to be understood about possible quantum-quantum decohering-classical propagating organization of processes. See Alicia Juarrero, *Dynamics in Action* (Cambridge, Mass.: MIT Press, 1999) for a discussion of boundary conditions and intention. Perhaps we can, in fact, find our way to a responsible free will.

CHAPTER 15

246 *the split between reason and the rest of human sensibilities:* C. P. Snow, *The Two Cultures and the Scientific Revolution* (New York: Cambridge University Press, 1959).

248 *Christian Bok, a prize-winning Canadian poet*: I thank Christian warmly for many wonderful discussions, not to mention dinners shared. Christian Bok, personal communications, 2006–2007.

254 *The two elements of cultures need not be divided:* I find myself increasingly trying to understand what it might be that's necessary to reunite our full humanity. Perhaps with some amusement, I am, as a scientist with some forty years of science behind me, taking courses on playwriting. I want to report to the reader something I am finding that is wonderful. The mode of creativity in scientific discovery feels deeply similar to that in the discovery of what one's characters are, and do. In neither case, it seems, do we, the scientists and the writers, let alone artists and composers, know beforehand what we will create. Of course, I am not the first to say this. However, it is one thing to reason about this in the third person. It is quite another to find it happening in oneself as one tries to span Snow's two cultures.

CHAPTER 16

258 *Paul Hawken:* Paul Hawken, *Blessed Unrest: How the Largest Movement in the World Came into Being and Why No One Saw It Coming* (New York: Penguin Group, 2007).

CHAPTER 17

261   *Frans de Waal describes this in* Good Natured: Frans de Waal, *Good Natured:
        The Origins of Right and Wrong in Humans and Other Animal* (Cambridge,
        Mass.: Harvard University Press, 1996).
268   *But as Harris points out, those goods included the neighbor's slaves:* Harris, S.
        Letter to a Christian Nation. New York: Alfred A. Knopf, 2006.
269   *During the period when Cromwell had replaced English royalty:* I thank Robert
        McCue for this history, personal communication, 2006.

CHAPTER 18

273   *I was honored to be invited to a "charette":* I thank Paula Gordon for inviting me
        to this event, and thank its participants warmly.
275   *In* Fast Food Nation, *Eric Schlosser:* Eric Schlosser, *Fast Food Nation: The Dark
        Side of the All-American Meal* (New York: Houghton Mifflin, 2001).
277   *As Jared Diamond remarks:* Jared Diamond, *Collapse: How Societies Choose to
        Fail or Succeed* (New York: Viking Penguin, 2004).

CHAPTER 19

283   *a sense of God as the creativity in the universe:* The Harvard theologian, Gordon
        Kaufman, has written on this topic extensively. See G. D. Kaufman, *The Face
        of Mystery: A Constructive Theology* (Cambridge, Mass.: Harvard University
        Press, 1995).

BIBLIOGRAPHY

Anderson, Philip W. "More Is Different," *Science* 177 (1972): 393.

Clayton, Philip. *Mind and Emergence: From Quantum to Consciousness*. Oxford, U.K.: Oxford University Press, 2004.

Collins, Francis S. *The Language of God: A Scientist Presents Evidence for Belief.* New York: Free Press, 2006.

Crick, Francis. *The Astonishing Hypothesis: The Scientific Search for the Soul.* New York: Simon and Schuster, 1994.

Davidson, Donald. *Essays on Actions and Events*. Oxford, U.K.: Clarendon Press, 2001.

Dawkins, Richard. *The God Delusion*. New York: Houghton Mifflin Company, 2006.

Dennett, Daniel. C. *Breaking the Spell: Religion as a Natural Phenomenon*. New York: Viking, Penguin Group, 2006.

Flanagan, Owen. *The Really Hard Problem: Meaning in a Material World*. Cambridge, Mass.: A Bradford Book, MIT Press, 2007.

Goodenough, Ursula. *The Sacred Depths of Nature*. New York: Oxford University Press, 1998.

Harris, Sam. *Letter to a Christian Nation*. New York: Alfred A. Knopf, 2006.

Hawken, Paul. *Blessed Unrest: How the Largest Movement in the World Came into Being and Why No One Saw It Coming*. New York: Penguin, 2007.

Juarrero, Alicia. *Dynamic in Action: Intentional Behavior as a Complex System*. Cambridge, Mass.: A Bradford Book, MIT Press, 1999.

Kauffman, Stuart. *Investigations*. New York: Oxford University Press, 2000.

————. *At Home in the Universe: The Search for Laws of Complexity*. New York: Oxford University Press, 1995.

————. *Origins of Order: Self Organization and Selection in Evolution*. New York: Oxford University Press, 1993.

Kaufman, Gordon. *In the Beginning, Creativity.* Minneapolis, Minn.: Augsberg Fortress, 2004.

————. *The Face of Mystery: A Constructive Theology.* Cambridge, Mass.: Harvard University Press, 1995.

Koch, Christof. *The Quest for Consciousness: A Neurobiological Approach.* Englewood, Colo.: Roberts and Company, 2004.

Laughlin, Robert. *A Different Universe: The Universe from the Bottom Down.* New York: Basic Books, 2005.

Lewis, Bernard. *What Went Wrong: The Clash Between Islam and Modernity in the Middle East.* New York: Harper Collins Perennial, 2002.

Rosen, Robert. *Life Itself: A Comprehensive Inquiry into the Nature, Origin and Fabrication of Life.* New York: Columbia University Press, 1991.

Smolin, Lee. *The Trouble with Physics: The Rise of String Theory, the Fall of a Science, and What Comes Next.* New York: Houghton Mifflin Company, 2006.

Steiglitz, Joseph. *Globalization and Its Discontents.* New York: W.W. Norton and Company, 2002.

Susskind, Leonard. *The Cosmic Landscape: String Theory and the Illusion of Intelligent Design.* New York: Little, Brown and Company, 2006.

Thagard, Paul, ed. *Mind Readings.* Cambridge, Mass.: A Bradford Book, MIT Press, 2000.

Unger, Roberto Mangabeira. *Social Theory: Its Situation and Its Task.* New York: Verso, 2004.

Weinberg, Stephen. *Dreams of a Final Theory: The Scientist's Search for the Ultimate Laws of Nature.* New York: Random House, 1994.

————. *The First Three Minutes: A Modern View of the Origin of the Universe.* New York: Basic Books, 1977.

INDEX

Made in United States
North Haven, CT
28 October 2021